THE DIAGNOSTICIAN

AMERICAN HEALTHCARE IS SICK, MISDIAGNOSED, AND GETTING WORSE; IT'S TIME TO CALL THE DOCTOR

THOMAS POWELL, MD MS

The clinical case vignettes presented in this book are composite descriptions drawn from the author's professional experience. Identifying details have been altered to preserve patient and organizational privacy. Quoted statistics, references, and citations are intended for educational and illustrative use; they are not a substitute for individualized clinical, legal, or financial advice. The imagery is produced from iteratively constructed context-based AI prompts, copying nothing from other works.

First Edition, 2026.

Published by CMIO LLC Publishing

Hardcover ISBN: 979-8-9961314-0-2

Paperback ISBN: 979-8-9961314-1-9

eBook ISBN: 979-8-9961314-2-6

Library of Congress Control Number: 2026912541

To my parents, Tommy and Kate; who listened to my ramblings then, and who I know still listen now, one from a little further away.
To my sisters, Kathleen and Michelle, and my brother Rob; who have always made me laugh, and whose laughter I still hear.
To my wife, Cassandra, who has put up with my behavior for way longer than expected, and has chosen to stay for more.
To my kids, David, Jenna, and Danielle, who make me proud every day.
To Sloane, and my expanding family.

FOREWORD

BY DAVID LUBARSKY, MD, MBA

David A. Lubarsky, MD, MBA, FASA, is President and Chief Executive Officer of Westchester Medical Center Health Network (WMCHealth), an 1,800-bed, nine-hospital academic health system serving the Hudson Valley of New York. He previously served as Vice Chancellor of Human Health Sciences and CEO of UC Davis Health, a $5.5 billion enterprise spanning a top-ranked medical school, children's hospital, and academic medical center, where he built one of the nation's earliest telehealth backbones and advanced AI-enabled care. Over seventeen years at the University of Miami, he was founding CEO of an 1,100-physician group practice and chief medical officer of a $2 billion health system. Trained as an anesthesiologist with a cardiovascular fellowship at NYU, he holds an MD from Washington University in St. Louis and an MBA from Duke's Fuqua School of Business, where he graduated as class valedictorian. One of the most-published practicing anesthesiologists in the United States, with more than 140 peer-reviewed papers, he has been named one of Modern Healthcare's 100 Most Influential People in Healthcare and Press Ganey's CEO of the Year.

BY DAVID LUBARSKY, MD, MBA

I have spent my career on both sides of a divide that should never have existed; practicing and teaching medicine on one side and running the health systems that deliver it on the other. I developed as an anesthesiologist, earned an MBA because I wanted to understand the economics that govern care as well as the methodology that governs healthcare, and have since had the privilege of leading large health systems on both coasts. From that vantage point I have read a great many books about what is wrong with American healthcare. Most either diagnose without offering a remedy or prescribe without making a diagnosis. The Diagnostician does neither. It is the rare book that takes a history, examines the patient, reads the data honestly, and reasons its way to a treatment. It is rarer still in being right.

Dr. Powell's central move is deceptively simple: he applies the discipline of the clinician to the institution itself. Physicians know that you do not treat a fever without trying to understand its cause, and yet that is very nearly how we have run healthcare transformation for twenty years; buying technology, launching initiatives, and chasing metrics in place of a diagnosis. He names that pattern for what it is, then does the harder work of identifying the underlying disease and who is positioned to treat it.

As someone who has spent a great deal of my career on the economics of care, I was struck by how clearly he ties data to outcomes to decisions with what he calls the data proposition, and how unsentimentally he follows the money. His account of the mismatch between the severity of a patient's illness and the way that care is documented and reimbursed is exactly right, and it is no abstraction: it is often the difference between a rural hospital that survives and one that closes. He understands that real solutions must account for genuine costs and genuine savings, and for every party to the problem rather than a favored few.

The spine of the book is its argument for the executive-physician; the leader able to hold clinical medicine, information technology, workflow, and executive judgment in a single mind, and who is genuinely chartered to synthesize them. I will not pretend this is an easy case to put to a sitting chief executive; it asks the C-suite to concede that some problems cannot be solved from a purely administrative chair. But Dr. Powell is both correct and fair. He does not indict leaders who are trying; he describes a capability most organizations have simply never built, and offers a blueprint for building it. The leaders I most admire are precisely the ones who will recognize themselves in these pages and be spurred rather than stung.

On technology he is at once enthusiast and skeptic, which is the only honest posture left to us. He is candid about how our electronic records can hide data, bury clinicians in noise, and tolerate poor data quality; clear-eyed about why so few clinical decision support tools are trusted and used; and his framework for governing artificial intelligence, asking of every algorithm its indication, its dose, its monitoring, and its stop rule, as we would of any new drug, is the most sensible approach to AI in medicine I have encountered. Having helped build telehealth and AI-enabled care at scale, I can attest that the absence of exactly that discipline is where the harm occurs.

What finally sets this book apart is that it is usable. The questions at the close of each chapter let a reader hold a real institution up against the standard the author proposes, and the closing chapters offer not a lament but a plan. In a field that changes faster than any clinician can keep pace with, one that increasingly makes the case for a continuously learning health system, that combination of rigor and practicality is precisely what is needed. This is a book a board can act on.

I recommend it without reservation to clinicians, to the executives and trustees who govern our systems, and to the engineers and data scientists who build our tools. Dr. Powell writes as a clinician and an informaticist who plainly knows his subject, and he has produced something unusual: a diagnosis of American healthcare rigorous enough to be believed and constructive enough to be followed. Recognition of the illness is the beginning of healing, and this is a very good place to begin.

BY DAVID LUBARSKY, MD, MBA

David A. Lubarsky, MD, MBA, FASA

President and Chief Executive Officer Westchester Medical Center Health Network (WMCHealth) Valhalla, New York

FOREWORD

BY ED HAMMOND, PHD

W. Ed Hammond has spent five decades at the center of American health informatics. Among the earliest architects of electronic health record system standards in the United States, he has contributed foundational work in clinical data standards, health information exchange, and the application of biostatistics to clinical decision-making. He has published extensively, served in national advisory roles, and led the development of one of the country's premier clinical informatics training programs. He is Director of the Duke Center for Health Informatics, Director of the Management Masters in Clinical Informatics, and Professor in the Departments of Family Medicine and Community Health and Biostatistics and Bioinformatics at Duke University School of Medicine.

This book is one of the best I have read. It is easy to read and understand. Dr. Powell writes from the perspective of a clinician and an informaticist, and he clearly knows what he is talking about. He uses a novel approach to defining and understanding what is wrong with the current healthcare system, identifying both the nature of the problems and who is best positioned to solve them. He examines the cost of current approaches and explains clearly why they are not solving, and will not solve, the underlying problems.

He appropriately discusses why the C-suite cannot solve these problems. C-suite leaders have an entirely different focus, and the problems of the current system do not directly affect them. The author has been in the trenches. He knows the difference.

Healthcare is changing rapidly. New knowledge obsoletes existing knowledge, and the question of how practicing clinicians stay current has no easy answer. We need continuous Learning Health Systems. Dr. Powell identifies the many problems of the current electronic health record (EHR) systems: they hide data, they saturate users with unneeded information, and data quality remains unacceptable. Clinical decision support (CDS) systems are important and useful when deployed correctly. The author notes that few CDS algorithms are currently in use, and then asks why.

Dr. Powell addresses the importance of attending to the chief complaint and ensuring it is resolved. It is a straightforward clinical principle that is routinely violated at the system level.

One of the strengths of this book is the set of questions at the end of each chapter, which allows readers to evaluate their current systems against what the author proposes. He ties data to outcomes and then to decision-making, which is an important step in understanding the significance of high-quality data, and what he calls the data proposition.

Solutions must include all people involved in the problem and must account for real costs and savings to the healthcare system. An important observation ties reimbursement to severity of illness: the mismatch between clinical care and billing is pervasive, and if taken seriously, would prevent many small rural hospitals from closing due to financial failures. Most problem lists are not kept current and have limited clinical value. Problems that have been resolved need to be documented as such, not deleted.

The author impressively addresses multiple dimensions of healthcare informatics and presents a coherent argument across them. At a time when informatics is disappearing from many health systems, replaced by the terms “digital health” or “data science,” his discussion is extremely important. There is now a certification process for clinical informatics. Dr. Powell proudly labels himself an informaticist, and this book makes clear why that designation matters.

The future of health systems must integrate academic skills across multiple disciplines: clinical medicine, computer science, artificial intelligence, biomedical informatics, and statistics. Problems should be approached by interdisciplinary teams, not by individuals working in isolation.

Artificial intelligence and its role in future health systems receive treatment here as well. AI systems are growing more capable rapidly, and they will play an increasingly important role in healthcare. It is a technology this book’s framework is well positioned to govern.

The author has done an outstanding job engaging the relevant literature. He has not missed significant advancements related to his topic, and his references provide excellent supplemental reading for anyone wishing to go deeper.

This is an outstanding book, and I strongly recommend its widespread reading by anyone involved in healthcare; C-suite leaders, clinicians, developers, bioinformaticists, and data scientists.

W. Ed Hammond, PhD
Director, Duke Center for Health Informatics
Director, Management Masters in Clinical Informatics
Professor, Department of Family Medicine and Community Health
Professor, Department of Biostatistics and Bioinformatics
Duke University School of Medicine

PREFACE

AN OPENING WORD FROM THE DIAGNOSTICIAN

This book comes from both my passion and concern, motivated by a career watching the slow deterioration of the American healthcare system. Not from the outside, but from the inside, in the rooms where the decisions get made.

I have spent my career at the intersection of clinical medicine, health information technology, clinical engineering, and organizational leadership. My career spans academic, community, pediatric, and integrated delivery systems. In most of them I was the first, and only, physician asking certain difficult questions out loud.

What I found is no mystery. The problems we have are not only large, they are largely well known. They are documented in the literature and confirmed by every clinician who works a shift. Remarkably, few of them have ever been addressed. They persist not because they resist treatment but because the case has never been referred for treatment. Most of them respond, and they respond to a discipline nearly every organization has heard of but have never empowered or staffed properly.

The discipline is *applied clinical informatics*. That work, and the leadership it demands, is this book's subject.

I wrote this book for *everyone*, because all of us will eventually be the patient, or the family member beside the patient, affected by decisions made in rooms we will never see.

If you are in a roll that sets *healthcare policy*, this book describes what those rules have produced at the bedside, and the diagnostic steps that were skipped when the country committed tens of billions of dollars to a treatment it had never tested.

If you sit on a *hospital board*, this book contains the questions about people and outcomes that belong in your monthly packet and are almost certainly missing.

If you run *hospital operations*, this book describes the blind spots that keep your dashboards green until the problem surfaces in a safety report, a financial statement, or a regulatory finding.

If you are a *physician* who is tired, frustrated, disengaged, or planning your exit, this book examines who broke your clinical workday and how. We were invited into some of those rooms, and we may have inadvertently signed the consent form.

If you are a *nurse, pharmacist, advanced practice provider, or an allied health professional* working at the edge of your license with less time, less support, and fewer tools than the work requires, the same accounting applies to you.

And, if you work in *informatics*, my own discipline, you will find a thorough treatment here, and maybe the most uncomfortable one. We are a large part of the answer. Currently, we are also a substantial part of the problem.

A NOTE ON METHOD AND STRUCTURE

My method is not novel in form, just in its focus. Every physician already knows how to use it. This book follows the *clinical method*, and the structure of the argument is the structure of a workup. Parts I and II take the history and run the tests. Part III examines the examiner. Part IV commits the diagnosis in writing. Parts V and VI prescribe and monitor. Part VII refers the patient onward, to a country that has not yet accepted the case. (See Figure 0.0.)

THE ARC OF THE BOOK

American Healthcare's Workup

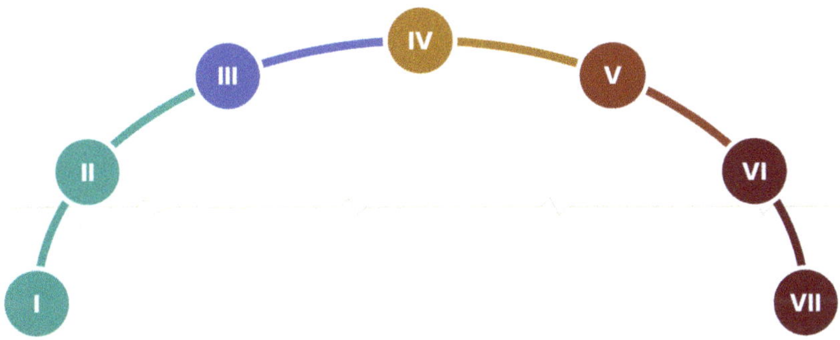

THE DISCHARGE PLAN

Discharge to the care of the Diagnostician and clinical informatics.

FIGURE 0.0

The clinical method for American Healthcare.

I have practiced it that way for decades, and it consistently works.

The hard work is giving you the numbers.

The results exist; I will show them to you. Interestingly, in my whole career, none of them have ever been requested. For most locations, the function's objectives and metrics were never defined, so there was no standard to hold informatics to. This book establishes both. By the last page you will be able to hold me and any other person who claims to do this work, to a solid function and number.

HOW THIS BOOK EVALUATES A SYMPTOM

Every finding is examined five times.

First, what it is.

The lesion itself, described as a physician would describe it, with the evidence that establishes it.

Second, what issues it causes.

No failure in a healthcare organization stays where it started. Documentation burden does not remain documentation burden. It becomes cognitive load, then after-hours work, then disengagement, then a resignation letter, then a vacancy, then agency staffing, then a ratio that makes the next resignation more likely. A missing data element does not remain a missing data element. It becomes a denied claim, a quality measure abstracted by hand, a case-mix index that reports the patients were healthier than they were, and eventually a payment that does not reflect the care that was delivered.

This is the part of the disease that has been consistently missed, and there is a structural reason for it.

The money almost never surfaces where the damage begins.

The chief financial officer sees the cascade and cannot trace it back. The chief information officer sees the lesion and cannot follow it forward. Walking that chain in both directions requires a person trained in clinical practice and in the systems at the same time, and that person is rarely in the room.

Third, what it costs.

Every cascade in this book has a price. I have put one on each of them using published sources, shown in full, taking the conservative end of every range. Those figures accumulate quietly through the first seven chapters. In Chapter 8 they are added together, which as far as I can establish has never been done before in this field.

Fourth, what its non-monetary impact is.

Some of what this disease costs Americans cannot be converted into dollars. I have declined to price them. I would hope that the reader agrees that the argument does not need them.

Fifth, what can be done about it.

Can a properly trained Diagnostician treat healthcare's illness? And, does it lead to partial healing or a genuine cure?

Every dollar figure in this book comes from a published source, and the sources are named. Where a range exists, I have used the low end. Where I could not verify a validated figure, a good faith estimate is provided.

The reader is welcome to keep a running total.

THE BEDSIDE QUESTIONS

Every chapter closes with Bedside Questions. They are a focused history and examination the reader can conduct on their own organization. Each one has a factual answer that exists in their own system. An executive who gets a favorable or unfavorable answer has learned something.

A NOTE ON TERMINOLOGY

Applied clinical informatics is the discipline that integrates clinical reasoning, technology architecture, workflow engineering, data analytics, and executive governance in the modern clinical environment.
Data proposition is the chain of information that runs from an observation at the bedside, through capture, mapping, aggregation, and interpretation, to a decision about a patient, and ends as an operational outcome. Every link can fail, and an organization's data are only as good as their weakest link.
Executive-physician is the Diagnostician. They are the physician attending to the whole organization the way a lead physician attends to a whole patient. Reading everything, integrating everything, and answerable for whether it gets better. The role has three parts, and they are not the same thing. **The Name** is what appears on the organizational chart. **The Form** is the chair in the C-suite, the defined authority, the budget, and the team. **The Function** is the work itself, which becomes possible only when the first two are real.
Governance is coordination of decision-making. Whether clinical or organizational. Every relevant discipline is present. The data are read together. One plan binds.

Organizations use different names for it. Chief medical information officer, chief health information officer, chief digital officer, and in a great many places no name at all. In my case the name has been CMIO. The Name is not the argument.

Most American organizations have the Name. Fewer have the Form. Very few have the Function, and the distance between them is what Part III is about.

The Diagnostician is what this book calls that physician, and the title describes the method rather than the Name. Take the history, examine the patient, consider the differential diagnoses, run the workup, commit to a diagnosis in writing, prescribe against it, and measure the response. Applied to an organization instead of a person.

Citations follow the author and date. The bibliography gives complete information for every work cited, and the glossary covers abbreviations. All clinical vignettes are composites, built from patterns that recur across many engagements, anonymized to protect identity, and faithful to the diagnostic lesson in each.

Let us begin the workup.

CONTENTS

INTRODUCTION

NAMING HEALTHCARE'S DISEASE

HEALTHCARE IS SICK: EXAMINING THE PATIENT

WHEN A PATIENT IS ADMITTED WITH A COMPLEX, MULTI-SYSTEM ILLNESS, a *lead physician* is assigned to the case. Every consulting physician who assists the case is important in their own right; only one of them is the lead. The lead takes the complete history. The lead performs an examination. The lead integrates the labs, the imaging, the consults, and the bedside findings into a single coherent picture, makes the diagnosis, and writes the orders. And, unlike the specialists or the discharge planner, the lead is held accountable when the patient gets better or gets worse.

Now imagine the same patient with heart failure, kidney injury, pneumonia, delirium, a non-healing wound, and a low-grade fever. Admitted with no lead assigned. Each consultant arrives, addresses their own organ system, leaves a note, and then goes home. The cardiologist increases the diuretic. The nephrologist warns it will worsen the kidneys. The pulmonologist starts an antibiotic. The hospitalist on shift turns over twice in 24 hours. The entire chart has never been read. The clinicians have never assembled the complete picture. The patient has never been examined as a whole organism.

The orders accumulate. The interventions multiply. The patient, buried under this divided attention, gets steadily worse.

This is the patient we will be discussing. The patient is our American healthcare system.

FOR TWO DECADES, THE PRESCRIPTION WRITTEN FOR THIS PATIENT HAS been the same: more government regulation, more financial incentives, more quality measures, more monitoring, more penalties, all requiring more extraneous data, which ultimately requires more technology. A larger, more sophisticated EHR. A new industry analytics platform. A better patient portal. The latest AI tool.

They all arrived without much understanding of the systems that they would impact, and each arrived with a disruptive implementation, a training program, and a press release. The cost has been extraordinary, and the total has never been summed. With rare exceptions, the trajectory of the healthcare disease has not changed.

A patient this sick does not need another specialty-focused consultation. Every consultant in the building has already been called, and has left a note. What this patient needs is a Diagnostician. This book is about the leader who can change that trajectory: the Diagnostician who treats the organization itself as the patient.

THE PATIENT IS WORSENING

The numbers tell the story. In 2024, the United States spent **$5.3 trillion** on healthcare, 18% of gross domestic product (GDP), the highest health expenditure, by number and share, of any country in human history (CMS, 2025c). The actuaries project that the share will keep climbing through the next decade (Keehan et al., 2025).

For that money, Americans receive measurably worse outcomes than peers in other high-income countries. Life expectancy is lower.

Maternal mortality is higher. Avoidable death is higher, a growing reality that the World Health Organization now tracks as a leading cause of global harm (WHO, 2024).

Estimates of death from medical error in American hospitals range from 44,000 to more than 250,000 a year, depending on the methodology (Kohn et al., 1999; James, 2013; Makary and Daniel, 2016).

Diagnostic error is astoundingly common. Newman-Toker and colleagues estimate that 795,000 Americans are permanently disabled or die every year because of a misdiagnosis (Newman-Toker et al., 2024).

These figures use different methods, count different outcomes, in different settings. Read together, they describe a system that harms people at scale and none of the facts are in dispute. The case needs to be referred for a full workup.

The clinicians delivering this care are exhausted. A survey of more than 5,000 U.S. physicians put the median EHR rating at 45.9 on the 100-point System Usability Scale, a grade of "F" (Melnick et al., 2020). For every hour of direct patient contact, ambulatory physicians spend nearly two *additional hours* on the EHR (Sinsky et al., 2016).

Burnout now affects more than 40% of U.S. physicians, a rate that has persisted at crisis levels for more than a decade (AMA, 2024). This burnout is contributing to a predicted U.S. physician shortage of up to 86,000 by 2036 (AAMC, 2024).

The financial picture is no kinder. Median hospital operating margins remain razor thin. Kaufman Hall's median operating margin for full-year 2025 stood at just 2.9%, and roughly 2 in 5 are operating at a loss (Kaufman Hall, 2026).

Reimbursement, increasingly contingent on documented quality, is becoming a function of how the EHR is configured, how data flow from the bedside to the dashboard, and how reliably clinical workflows produce the evidence that payers require.

Each of these findings has a price; the patient has not been examined, and every therapy to date has been aimed at nothing specific. By price, I am not referring to the cost of American healthcare, which is well established. Many can quote that number. *The cost of this specific illness*, the one this book is about, is in lives, health, and financial waste that a trained Diagnostician can reach.

The numbers exist, have never been assembled before, and are significantly more than what it costs to put a qualified Diagnostician in the chair. The rest of this book addresses who should be doing it, and what they should be doing. It includes the arithmetic of return on that investment.

THESE ARE the symptoms of your patient. These are not isolated complaints. Together, they form a recognizable syndrome.

The patient is sick, getting worse, and running the same therapies that have been failing for two decades.

The issue is not that the medications don't work or do not exist. The issue is that *the patient has not been examined,* and every therapy to date has been aimed at nothing specific. We are spending the time and money to treat a disease we can't identify. (See Figure 0.1.)

ADMISSION SUMMARY · 2026

American Healthcare

Eight vital signs, every year. Every one abnormal.

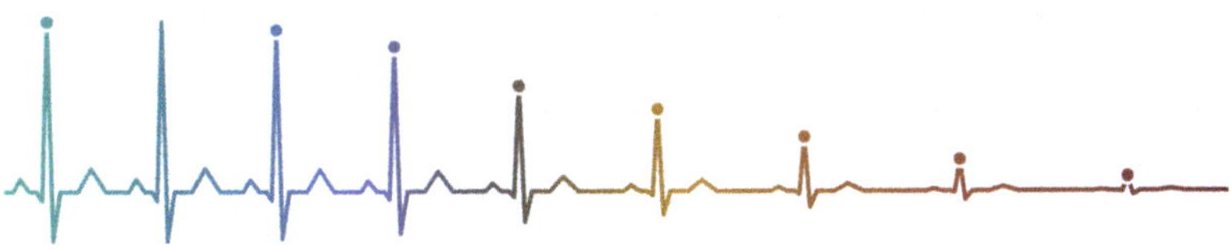

HEALTHCARE SPENDING, YEARLY	SERIOUS HARM FROM DIAGNOSTIC ERROR	ADMINISTRATIVE COMPLEXITY, YEARLY	HOSPITAL MEDIAN MARGIN
$5.3T	**795K**	**$265.6B**	**2.9%**
CMS, 2025c	Newman-Toker et al., 2024	Shrank et al., 2019	Kaufman Hall, 2026

HOSPITALS OPERATING AT A LOSS	NURSING TURNOVER, YEARLY	LOOMING PHYSICIAN SHORTAGE	PHYSICIAN BURNOUT
2 in 5	**17.6%**	**86K** projected by 2036	**40%**
Kaufman Hall, 2026	NSI, 2026	AAMC, 2024	AMA, 2024

DIAGNOSTIC IMPRESSION

Multi-system illness

RECOMMENDATION

Assign a lead physician

Eight vital signs, one patient. None of them heal by adding another vendor.

FIGURE 0.1

The symptom profile of American healthcare, measured every year.

THE WRONG TREATMENT FOR THE DISEASE

Medicine has a name for treatments prescribed without a clear diagnosis. They are called *empiric therapy*, and they have a real place; when the patient is septic and cultures will take 48 hours to identify the specific bacteria, you cannot wait. You start broad-spectrum antibiotics and you narrow when the data return.

But empiric therapy that is never narrowed becomes something else. It becomes polypharmacy. Drug interactions. Adverse effects that are themselves treated empirically, with new medications that produce new adverse effects that demand new empiric trials, until the patient is on a dozen drugs that were each defensible in isolation and collectively are causing harm. Geriatricians have a name for this too. They call it *the prescribing cascade*, and they consider it a hallmark of bad medicine.

For two decades, American healthcare has been on a prescribing cascade.

The original prescription for America was rational on the surface. In 2009, the Health Information Technology for Economic and Clinical Health (HITECH) Act authorized $32 billion to accelerate adoption of electronic health records (EHR) (Blumenthal, 2010; Buntin et al., 2010). The diagnosis underlying it, that paper records were a structural cause of fragmented, error-prone, high-cost care, was largely correct. The treatment achieved its near-term goal: by the early 2020s, virtually every U.S. hospital and most ambulatory practices had adopted certified EHR technology (Adler-Milstein et al., 2017; Washington et al., 2017).

Then the cascade began.

When EHR adoption alone did not improve outcomes, the prescription became meaningful use. When meaningful use produced documentation rather than better care, the prescription became value-based purchasing. When value-based purchasing exposed how poorly the data moved and were largely ignored, the prescription became more interoperability. When interoperability did not by itself improve outcomes, the patient portal arrived. And, as each treatment failed to bend the curve, the prescription escalated to more technology, including artificial intelligence.

Every treatment was thought of as rational. Not one of them was preceded by an adult conversation and a diagnostic workup of the flailing organizations themselves.

The wards were not walked. The history was never taken. The patient was never examined. And, the patient kept getting worse on increasing empiric therapy.

American healthcare has been operating in a state of clinical insanity, doing the same things day after day and expecting better results, for over seventeen years. Multiple escalations. Tens of billions of dollars. The same diagnostic steps skipped every time. This book asks American healthcare to stop recycling insanity, makes the argument for how it can, and offers the method that should replace it.

The pathway that American legislators chose to pursue, concerning healthcare's issues, was not lacking expert instruction and warning. Nearly all of it was unsolicited and unheeded. The scientists who founded the clinical informatics field wrote down *what a medical record is for*, in print, decades before the money moved.

Ledley and Lusted explored in 1959 *how physicians reason*, a handy thing to know when designing technology used for reasoning. Weed explored in 1968 *what information the medical record should provide* to the person keeping it, a handy thing to consider as well. Slack explored in 1977 *who owns their medical record*. An Institute of Medicine committee explored in 1991 *whether the record should follow the patient across entities and encounters*. The last two studies recognized the longitudinal record as the *patient's record,* not an institutional document. Hammond explored in 2005 the need for *basic data standards* for records that will be shared.

Those questions were all asked openly before the country spent a dollar on HITECH and "meaningful use." Most of them are still unanswered to this day. Chapter 17 goes back and reads those notes, because a physician who does not read the old chart is missing some key history.

The entire country is predominantly using proprietary EHR technology, controlled by a corporate business case, and not directed at healthcare's therapeutic case. The largest indictment of this approach was published more than a decade ago and has been confirmed at every checkpoint since. In 2005, RAND researchers projected that aggressive adoption of health IT would save roughly $81 billion a year. In 2013, the same group revisited those projections and found the savings had not materialized. None of it.

The failure was not technological, because the software was turned on and it worked. What was missing was the organizational redesign required to convert capability into reality (Kellermann and Jones, 2013). (See Figure 0.2.)

THE CENTRAL ARGUMENT

The Diagnostic Approach to a Sick Organization

Seven steps, repeated, until the organization recovers.

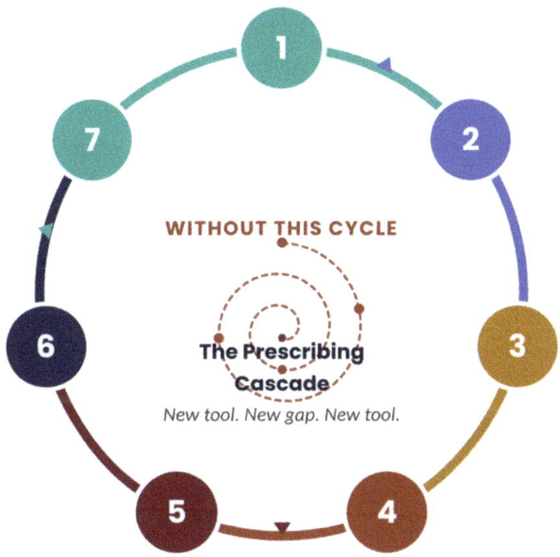

1 Take the Full History
Clinical, operational, and financial.

2 Examine the Patient
Vital signs, read as a system.

3 Consider the Differential
Symptom vs. underlying disease.

4 Order the Right Workup
Audit the data. Watch the workflow.

5 Commit to the Diagnosis
Write it down. No hedging.

6 Write the Treatment Plan
Specific. Measurable. Owned.

7 Measure the Response
Implement. Monitor. Refine.

The disease is not the technology. It is the absence of a diagnosis.

FIGURE 0.2

The diagnostic cycle — the alternative to prescribing without a diagnosis.

WHAT IS WRONG WITH THE PATIENT

This is the diagnosis that most frequently gets missed. American healthcare is not suffering from a *technology deficiency*. It is suffering from a *diagnostic deficiency*. The technologies we have been purchasing for decades have never been aimed at a particular target. There has never been a Diagnostician at the bedside, no physician-leader synthesizing the whole picture, and no accountability for the comprehensive treatment plan.

We do not have an epidemic failure of intelligence or lack of effort in healthcare. We have a failure of *specialty-matching*. You do not send a brilliant orthopedic surgeon to read a complex abdominal CT scan. You do not ask the cardiologist to manage psychosis. The match is wrong, the training is wrong, and the tools of the trade are wrong.

Healthcare's executive suite has specialties too. And, they are struggling with *specialty-matching*. The available executive roles in modern healthcare have not been trained to treat the illness that the organization has. The chief information officer (CIO) is excellent at technology and is not trained to take a clinical history. The chief financial officer (CFO) is excellent at the balance sheet and is not trained to perform a workflow examination. The chief medical officer (CMO) is excellent at medical staff affairs and is not trained to govern the EHR build. The general counsel is excellent at assessing contract risk and is not trained to recognize a technology's clinical risk. The chief quality officer is excellent at root cause analysis and is not trained to determine whether the data feeding the quality report are sound. Not one of them has been built for the lead-physician role this case requires; none of them were presumed to be.

A consequential specialty mismatch started with the misguided assumption that the CIO, a competent professional in technology, could serve as an objective clinical resource during the most dramatic shift of the clinical environment in history. Chapter 9 examines what that assumption costs.

The correct specialty role is the one that applied clinical informatics provides and the Diagnostician holds. In most American healthcare organizations, this lead has either never been appointed, reports through another budget priority, has a name without the competencies, or has the name and skillset without the conditions needed to succeed.

This book is about why that must change, and what changes when it does.

HEALTHCARE AS THE PATIENT

Let me give you a way of seeing your organization that will be hard to unsee when we are done.

The patient is your healthcare organization.

A complex, interdependent, living system, with a chief complaint (declining margins, climbing burnout, deteriorating quality), a history (decades of mergers, consolidations, EHR migrations, leadership turnover), comorbidities (regulatory exposure, workforce attrition, payer pressure), and a worried family (your board, your community, the patients who depend on you).

The treatments are the technology.

Every EHR, every clinical decision support (CDS) rule, every AI model, every voice recognition platform, every telehealth pathway, and every interoperability engine your organization has ever licensed, configured, or deployed is simply a modality of treatment. Every modality has clear indications, contraindications, dosing requirements, expected effects, and potential adverse effects. Any of them can heal or harm, depending on the precision with which it is selected, prescribed, monitored, and adjusted. Given without a diagnosis, not one of them will reliably make the patient better.

The Diagnostician is the executive-physician.

The Form and Function are not replacements for the cardiologist (the CFO), the surgeon (the CIO), or the nephrologist (the CMO). Losing any of their specialty knowledge diminishes the treatment team. The executive-physician provides what no specialist can do on their own: synthesis of all clinical input into a single coherent plan for the whole patient, accepting accountability for whether the patient gets better.

The treatment plan is the evidence-based course of action.

A plan anchored in outcome measures. It targets the primary disease causing the symptoms rather than the symptoms themselves. And, it is constructed by a Diagnostician who has examined the patient, consulted with the specialists, and is willing to perform the ongoing follow-up.

This is a structural description of how healthcare succeeds or fails, and it is the same discipline whether the patient is a 69-year-old woman with chest pain or a 1,400-bed health system with stagnant quality scores. Once you see it, you will recognize the absence of it everywhere.

WHEN THE PATIENT IS NOT EXAMINED

The consequences of treating a patient without a diagnostic target are not theoretical. They have been measured in mortality, dollars, and organizational decline. This is never more observable than in the case of the American hospital system. Health IT, thought to be a benign therapy, is now a well-documented cause of patient harm: wrong-patient orders, medication errors built into order-entry design, critical results that surface in notification systems no physician reads, and alert override rates above 90%.

Poorly designed technology has been classified as an organization-level driver of clinician burnout, which is the national academy's own diagnosis. This is a systems failure, not a resilience deficit in the clinician (NASEM, 2019).

Documentation gaps cost hospitals tens of millions a year in revenue that was earned and never captured.

These findings will be developed, with evidence, in the chapters that follow. All of these outcomes were caused by the absence of a physician willing to examine the patient before prescribing the treatment. (See Figure 0.3.)

WHEN THE TREATMENT BECOMES THE DISEASE

Consequences of Technology Without a Diagnosis

Five patterns of ungoverned health IT, in the order the dose is delivered.

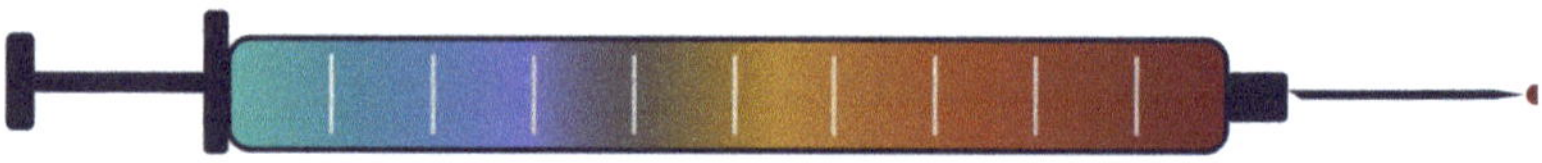

1	**Alert Fatigue**	**58 / 100K** wrong-patient order errors	**A simple verification prompt at order entry cuts this by 16%.** *Adelman et al., 2013*
2	**Burnout**	**45.9 / 100** EHR usability score — a grade of F	**Every 1-point gain in usability cuts burnout odds by 3%.** *Melnick et al., 2020*
3	**Revenue Leakage**	**$30K+** lost per clinician, per year	**Looks like a billing problem. It is bad documentation and corrupted data.** *Premier, 2025; Casalino et al., 2016*
4	**Regulatory Exposure**	**$1M** per information-blocking violation	**Purely informatics in origin, and it goes uncaught without the right leader.** *CMS, 2024; ONC, 2024*
5	**AI Harm**	**17.7% → 46.5%** Black patients flagged for care, before & after the fix	**The arithmetic was correct. The proxy was wrong.** *Obermeyer et al., 2019*

None of these failures is caused by the technology. Each one is caused by the absence of the physician-leader qualified to diagnose it and treat it.

FIGURE 0.3

The cost of treatment without diagnosis — five recognizable patterns.

WHAT THE DIAGNOSTICIAN MUST UNDERSTAND

The Diagnostician focuses on four competencies that usually operate in complete isolation.

Clinical Medicine

Not the textbook job description, but the knowledge and weight of a critically ill patient deteriorating at 3 a.m. The Diagnostician is a physician first, grounded in the fact that there are patients and families downstream of the technology and decision-making.

Information Technology

Not necessarily the ability to write code, but fluency in system architecture, interface specifications, the computational logic of rules, interoperability standards, and the EHR data structure. The Diagnostician must be able to ask the technical questions the organization is otherwise being told to accept on faith.

Workflow Engineering

Observing clinical care as it unfolds (at the bedside, at the workstation, in the OR, at the nursing station, in the clinic) and redesigning it so that the technology supports the work rather than impedes it. This is an engineering capability, applied to the clinical workflow. Understanding the difference between building high-quality versus unacceptable care processes into the workflow is the critical element.

Executive Leadership

Sitting at the leadership table as a peer, translating clinical priorities into technology decisions and technology capabilities into clinical impact. Governing, budgeting, measuring, and holding the line when a vendor pushes a solution not supported by a diagnosis.

This constellation of skills is not commonly taught. It is cultivated over years, across all four domains. That is why so few organizations have organically built the role, and why the profession that should be producing these leaders requires a deeper examination in Chapter 10. (See Figure 0.4.)

THE PROFESSION · WHO SHOULD LEAD

The Four Competencies of the Executive-Physician

No single pipeline produces all four. One chair must hold them all.

All four, in one chair, is rare enough that most health systems have never seen it.

FIGURE 0.4

The four competencies that meet at the executive-physician — each necessary, none alone sufficient.

WHAT RECOVERY LOOKS LIKE

The contrast, when it exists, is striking.

Some organizations have already committed to the Name, Form, and Function a skilled Diagnostician offers. You can identify these rare "zebras" because the role is seated in the C-suite with defined authority, resources, and mandates. They perform measurably better than organizations that have simply appointed an individual to the Name. The function is what predicts performance, not the title.

Successful intervention measured to date is tied to the applied informatics function, not an individual with a title.

Because this role lacks a clear, universal mandate, it remains understudied and difficult to compare. Individual organizations report successes internally, but a broad, published study of that data does not yet exist.

Those healthy organizations have a capacity and throughput that hold up through every flu season. The finance team trusts the data enough to plan. Technology investments are given defined objectives and held to post-implementation reviews. Safety events get analyzed and addressed, rather than filed. Their clinical decision support is governed and clinically relevant, with override rates defensible in public.

These examples are not from theoretical organizations; they exist today. These organizations understood early, through foresight or painful experience, that the patient needs a lead physician.

Their health can be shown in every dashboard metric healthcare is held to: quality scores, safety indicators, value-based purchasing performance (VBP), regulatory compliance, clinician engagement, operating margin. It also is shown by something no metric easily reaches. Clinicians inside those organizations stop describing their day as a war with the technology and start describing it as work the technology supports.

WHY THE WORKUP CANNOT WAIT

The need for the Diagnostician has never been more urgent, provoked by many converging forces.

The regulatory and payer environment is intensifying.

Medicare Access and CHIP Reauthorization Act of 2015 (MACRA), Merit-based Incentive Payment System (MIPS), value-based purchasing performance, information-blocking enforcement, interoperability mandates, the Trusted Exchange Framework and Common Agreement (TEFCA), and Centers for Medicare & Medicaid Services (CMS) Conditions of Participation, all increasingly require the installation of technology and the demonstrable use and impact (Burwell, 2015; ONC, 2023).

This is one gigantic list of words. It also shows the horrendous payer and regulatory gauntlet being navigated. Organizations without mature informatics leadership are falling behind and facing accelerating financial penalties. Savvy system boards and CFOs are asking harder questions like, “What is our technology spending buying?” Answering them is an informatics function.

AI has arrived in clinical care at scale.

The Food and Drug Administration (FDA) authorized 1,451 AI and machine-learning-enabled medical devices by the end of 2025, and 1,524 by March of 2026 (FDA, 2026). The shape of that curve matters most.

Generative and foundation-model devices have begun receiving market authorization, ambient documentation has moved from pilot to production across large parts of the country, and the interval between an AI model’s release and its appearance in a clinical workflow is now measured in months.

The governance capacity of American healthcare organizations has not changed significantly, or at all, in the same period. That gap is revisited in Chapter 13.

Clinical science, while lagging, has already raised concerns. AI deployed without governance has already been shown to produce harm, suggesting the need for more safeguards.

The workforce crisis is structural.

The projections for physician, clinician, and healthcare staffing have worsened. Hospitals are struggling to grow programs, while steadily losing their front-line clinical talent. Burnout is now classified as an occupational phenomenon by the World Health Organization, making stability even less likely.

It is becoming clear that technology engineered for the clinician becomes a retention tool, while technology that is not accelerates the leaving. *You cannot afford to burn out the workforce you cannot replace.*

Clinical complexity is increasing faster than can be absorbed.

Genomics, precision medicine, population health, digital therapeutics, remote monitoring, ambient documentation, and large-language-model-mediated clinical workflows are arriving at once. The organizations that will thrive in the next decade are doing the workup now (Shortliffe and Cimino, 2021).

THE TREATMENT PLAN AHEAD

The seven parts of this book will follow the clinical method mapped in the Preface, not as a borrowed metaphor but as an established skillset. By the time you reach the final page, you will not have merely read about the diagnostic method. You will have moved your organization through it.

The book's thesis is that American healthcare has a *diagnostic deficiency*. It has enormous quantities of information, extraordinary specialists in every domain, and no Diagnostician trained to read the whole picture objectively and diagnose. This diagnosis should be sought and always precede treatment.

Lastly, there is no universal therapy. Every organization is a different patient.

Built from two decades of evidence and direct experience, this book gives you the diagnostic vocabulary to name your organization's disease, the method to examine it rigorously, the case for the leader who can treat it, and, for the first time in this field, the return on investment numbers.

IT IS TIME TO GO TO WORK.

PART I

IDENTIFYING THE PRESENTING ILLNESS

Before the cardiologist orders a stress test or the nephrologist starts a diuretic, the lead physician has examined the patient. Before any treatment is prescribed, the lead physician is accountable for understanding what is wrong. Part I is the History and Physical of the modern American healthcare organization. Until you know what the patient is suffering from, every prescription is a guess.

CHAPTER 1

THE CHIEF COMPLAINT, HISTORY, AND PHYSICAL

CLINICAL SIGNS AND SYMPTOMS

THE CLINICAL ENCOUNTER IS THE MOMENT THAT DETERMINES WHETHER everything else in healthcare will succeed or fail. If the technology does not strengthen that moment, it weakens the outcome and the organization.

A PHYSICIAN WALKS INTO A PATIENT'S ROOM AT 7:14 IN THE MORNING.

The patient is a 69-year-old woman brought in overnight with chest pain. Vital signs are stable. Cardiac markers are trending down. The overnight EKG shows no new changes. She is sitting up, looking better than she did when she arrived.

The physician reviews what came in overnight, examines her,

listens to her, and begins the work that is the heart of medicine. The history, the exam, the labs, the imaging, and everything else known about this patient are pulled into a working diagnosis, a plan, and a series of decisions that will shape how she does.

That moment is where medical practice happens, and everything else in the building exists to support it.

Now consider what happens when that moment gets degraded. The EHR runs slowly. An alert that has nothing to do with this patient interrupts the cognitive flow. The critical lab result is hidden six clicks deep. The documentation template demands 15 minutes of typing about things that did not happen, so that the one thing that did can be recorded.

Each of these is survivable. Together they take time and attention away from the only work that matters, and the patient ultimately pays the price for it.

THE PATHOLOGY BEHIND THE SYMPTOMS

Every technology decision in a healthcare organization should start with a clinical question and check for clinical impact, even when it looks purely technical.

- What outcome are we trying to improve?
- What risk are we trying to reduce?
- What work are we trying to support?
- What evidence are we trying to put into practice?
- What will this do to the way people work with patients?

This seems self-evident and is rarely practiced.

In my experience, most healthcare-IT decisions come from somewhere else entirely. *Vendor capability*: the EHR can do this, so let's turn it on. *IT initiative*: the upgrade includes this, so let's turn it on. *Regulatory mandate*: CMS requires this, so let's turn it on. *Executive directive*: the chief executive officer (CEO) saw this demo at a conference, so let's get it and turn it on. *Competitive anxiety*: the hospital across town has this, so let's get it and turn it on.

None of these drivers contain a clinical question. There is a vendor pitch, a project deadline, a regulatory threat, an executive impulse, or simple peer pressure, and then a build proceeds, tied to no identifiable clinical priority.

The result is predictable, well documented, and now repeated in every health system in the country. Features get activated that no clinician uses. Configurations conflict with the way people work. Alert fatigue becomes pervasive. Order sets get assembled by IT analysts following vendor defaults rather than current clinical evidence (Ash et al., 2004). Documentation templates get built to capture billing rather than clinical thinking. What they produce is a long note that tells the next clinician almost nothing. The literature even has a name for it: *note bloat* (Weis and Levy, 2014; Rule et al., 2021).

THE CLINICAL ENCOUNTER, the reason the whole enterprise exists, is degraded rather than enhanced. This is the chief complaint.

The technology is usually capable of far more. What failed is the step before it. The problem the technology was bought to solve went undiagnosed, no target was set, no plan for treatment was built, and the prescription went out anyway.

No one took the history. No one examined the patient.

A word about the discipline this book rests on, because most readers have never encountered it as a discipline at all.

Applied clinical informatics is best understood by what it does, not only by how it is defined. It is the science of how medical information gets captured at the bedside, structured, moved, and put in front of the right person at the moment of decision. It is also the discipline that builds efficient, high-quality care delivery through technology and workflow, rather than around them. Clinical informatics is a recognized medical subspecialty with its own board certification, in the same way cardiology is a subspecialty of internal medicine. The physician who leads it inside a health system usually holds the title *chief medical information officer*, or CMIO. Whether that title comes with the authority and resources to practice the discipline is a separate question, and it is the one Part III takes up. Clinical informatics is not about the technology, it never was. It is the clinical discipline that ensures that the technology serves the clinical work, and answers for it when it does not.

THE CLINICAL SIGNS AND SYMPTOMS

A full examination of a healthcare organization's chief complaint turns up four groups of symptoms. Each has its own mechanism, its own evidence base, and its own diagnostic meaning. The work of the Diagnostician is to recognize them, describe them accurately, and connect them to a treatment plan.

Diagnostic Error and Patient Safety

Diagnostic errors affect roughly 12 million Americans a year in outpatient care, and about half of those have the potential for significant harm (Singh et al., 2014). In 2015, the National Academies of Sciences, Engineering, and Medicine's report concluded that most people would experience at least 1 diagnostic error in their lifetime, and that these errors drive a great deal of preventable death and disability (NASEM, 2015). The most recent national estimate puts the toll near 795,000 Americans killed or permanently disabled every year, with vascular events, infections, and cancers accounting for roughly 75% of cases (Newman-Toker et al., 2024).

Diagnosis is the process of reasoning under uncertainty, and it is among the hardest cognitive tasks any professional does. The physician gathers what data exist, forms hypotheses, weighs them against each other, and narrows toward an answer, knowing the whole time that the information is incomplete and the clock is running (Elstein et al., 1978; Croskerry, 2009).

Two kinds of thought run at once. One is fast and intuitive (*Type 1*), the pattern recognition that lets an experienced physician sense that something is wrong before being able to say why. The other is slow and deliberate (*Type 2*), the careful working through of possibilities and probabilities. Both are necessary, and both fail in well-characterized ways: anchoring on the first explanation that comes to mind, closing on an answer too early, being pulled by whatever case was seen most recently, or looking only for evidence that confirms what is already believed (Kahneman, 2011; Croskerry, 2003; Saposnik et al., 2016).

Software can help with all of those failures. Clinical decision support, usually shortened to *CDS*, is the layer of the EHR that puts information in front of a physician during the decision itself. It can surface the data that matter, raise an explanation that has not been considered, and flag a result that should stop some course of action (Graber et al., 2012; El-Kareh et al., 2013).

Kawamoto and colleagues identified four features that separate CDS that works from CDS that gets ignored.

- It must appear inside the work rather than requiring a separate step.
- It must recommend an action rather than merely report a finding.
- It must arrive at the time and place of the decision.
- And, the computer must generate and push it, rather than waiting for the clinician to go looking for it.

Systems doing all four improved clinical practice in 94% of trials (Kawamoto et al., 2005). Badly designed CDS does the reverse. It adds noise, scatters information across the screen, and adds weight to clinicians who are already doing as much as they can.

In 2004, Ash and colleagues traced exactly how health IT inadvertently manufactures new categories of error: rigid system logic that fights the way clinicians reason, interruptions that break concentration at the worst moment, and screen design that buries the one signal that matters underneath everything that does not (Ash et al., 2004).

This is self-induced iatrogenic disease at the level of thought. The treatment meant to support clinical reasoning is degrading it.

The modern patient safety movement began with the Institute of Medicine's 1999 report *To Err Is Human*, which estimated that preventable hospital error kills between 44,000 and 98,000 Americans a year (Kohn et al., 1999), a toll higher than the next five leading causes of accidental death combined. The number shocked the country. Later independent analyses suggest the real figure runs considerably higher (James, 2013; Makary and Daniel, 2016).

Health IT was supposed to be a large part of the cure, and where it has been implemented well, it has been. Computerized provider order entry (CPOE) has cut medication errors by between 48% and 86% in some locations (Bates et al., 1998; Radley et al., 2013).

Implemented poorly, the same technology causes harm. This is a finding not contested in the informatics literature.

Consider the simplest version. A physician has two patient charts open, because the work is made easier and quicker. An order goes into the wrong patient chart. Nothing on the screen makes the error obvious and nothing stops it. Adelman and colleagues measured how often this happens in a system without adequate identity verification and found roughly 58 wrong-patient orders per 100,000 in 2013, updated to as many as 91 in 2019. Then they added a single step, a prompt asking the physician to confirm the patient before the order goes through, and the rate fell by 16% (Adelman et al., 2013; Adelman et al., 2019). The example fix was not sophisticated. It was designed. The right Diagnostician can design the environment to help. They would also take on the question of whether the work requires two open charts, comparing the benefits to the risks.

The rest of the pattern is documented just as well. Koppel and colleagues found 22 separate ways a widely used order-entry system led to medication error, including fragmented displays, default doses that did not fit the clinical situation, and pharmacy stock driving substitutions that overrode what the prescriber intended (Koppel et al., 2005; Slight et al., 2016).

Singh and colleagues found that 6.8% of abnormal test results at two Veterans Health Administration (VA) facilities never received timely follow-up, with the design of the electronic notification identified as part of the cause (Singh et al., 2009).

Phansalkar and colleagues found that most drug-interaction alerts in commercial systems are of little clinical significance, which produces the "*boy-who-cried-wolf*" problem: physicians override everything, including the rare alert that would have mattered (Phansalkar et al., 2013). Override rates for drug-safety alerts run between 49% and 96% (Van der Sijs et al., 2006).

Regulators have been clear about the risks for more than a decade. The Joint Commission named health IT an institutional hazard in Sentinel Event Alert 54 and called for formal safety processes to be put in place (TJC, 2015). The Office of the National Coordinator for Health Information Technology (ONC) published the SAFER Guides so organizations could assess their own EHR safety in a structured way (ONC, 2025). ECRI has ranked health-technology hazards among its Top 10 Patient Safety Concerns every year since 2012 (ECRI Institute, 2024).

AND, THE PATIENT REMAINS UNTREATED.

Evidence-Based Practice

Balas and Boren estimated that newly discovered medical evidence takes an average of 17 years to reach routine practice (Balas and Boren, 2000; Morris et al., 2011). McGlynn and colleagues showed what that delay costs. Auditing more than 6,700 patients across 12 metropolitan areas, they found that American adults only received 54.9% of the care recommended for them (McGlynn et al., 2003). Two decades of work since has narrowed the gap in places without ever closing it.

This is a slow disease and a quiet one. Treatments that work go unused. Treatments that do not work stay in use. The waste is enormous, and almost none of it shows up on a report being read.

CDS is one of the most powerful tools available for closing that gap (Kawamoto et al., 2005; Bright et al., 2012; Osheroff et al., 2024). A well-built order set makes the right choice the default choice. Alert logic can put a guideline in front of someone at the moment of decision, assuming they are in the record at the decision-making point. Documentation can prompt the steps a protocol requires.

Deployed without governance, the same tools create noise instead of signal. For this to work, the Diagnostician must know the evidence, the workflow, the decision points, and how people behave under pressure. Wright and colleagues examined 23 healthcare organizations and found wide variation in how well CDS performed, with alerts overridden and ignored far more often where no formal governance existed (Wright et al., 2019; Cho et al., 2019).

Governance is the variable here, and it is also the hard work that goes unowned and underappreciated.

Care Coordination

Modern healthcare is now delivered by teams, spread across settings. At every transition, the patient's information must move accurately and completely. That has gotten harder.

Team dynamics and care environments are changing at a pace hard to match. Payment models have changed. Sub-specialties have narrowed. Resident duty hours have been capped. Workforces have shrunk. Each change required a reasonable response to isolated pressures. Together, they have multiplied the number of times a patient gets handed from one team to the next.

Every handoff is a chance to miss something.

Communication failure remains the most frequently documented root cause of sentinel events in American hospitals, and it concentrates at exactly these transition points. Discharge is the most fragile of them.

The industry has built considerable machinery in response. There are interoperability standards like Health Level Seven (HL7) FHIR, regional health information exchanges, and care-coordination modules inside the major EHRs. There are validated handoff tools like I-PASS, now built into digital dashboards.

None of these solves the problem, because this was never just a technology problem. The largest problems are workflow, how information gets organized, how they are governed, and how people behave when they are tired and busy (Vest and Gamm, 2010). Chapter 7 takes up the governance of it.

Clinical Documentation Integrity

The medical record is the data foundation of continuity, quality measurement, research, legal defense, regulatory compliance, and payment (Rosenbloom et al., 2010). When the record degrades through bloat, meaningless templates, and copy-paste, everything downstream degrades with it. The source gets corrupted, and everything drawing from it inherits the corruption.

Electronic health records are increasingly plagued by systemic duplication. Only 18% of the average inpatient progress note consists of manually entered text. The rest is compiled by copying previous entries or importing external data. The inpatient "bloat" impact is climbing from 33% of overall note volume in 2015 to 54.2% in 2020 (Wang et al., 2017; Steinkamp et al., 2022). Outpatient notes show the same trend in different numbers: 60% longer and 11% more repetitive across a single decade (Rule et al., 2021).

Meanwhile the thing that matters most goes missing. The physician's reasoning, what was suspected, and what was ruled out and why, are increasingly absent altogether or buried in pages of generated text that the next physician will never read.

The fix requires a Diagnostician who understands both what the record is for clinically and how to capture it technically: template design, structured data, natural language processing (NLP), voice recognition, ambient documentation, and how documentation integrity programs fit with all of it (Kreimeyer et al., 2017; Hodgson and Coiera, 2016).

FOUR DOMAINS, FOUR BODIES OF EVIDENCE, AND ONE REQUIREMENT running through all of them. Each domain demands a Diagnostician who understands them and their interactions: how clinicians actually think, how the software is built and can be modified, how clinical evidence is expressed as computable logic, and then how and where the decision happens. Then that someone has the authority to make the technology serve the work rather than the reverse.

None of that is contested in the literature. What is contested, inside most organizations, is *whose job it is*.

THE SYMPTOM CASCADE

When the chief complaint goes unexamined, the decline is not random. It follows a predictable path, and each step makes the next one harder and more expensive to reverse.

It starts with a *build decision* made without a clinical priority. An alert gets switched on, an order set changed, a template added, a workflow altered, none of it anchored to anything clinical. The load on the person at the bedside rises slightly. In any one moment the increase might be trivial. Multiply it across an 800-bed hospital, every clinician, every shift, month after month, and it stops being trivial.

Sustained workload above what a clinician can manage becomes a *safety risk*. High override rates are the visible sign. Wrong-patient orders keep happening where no verification step was built. Communication loops never close. Critical results keep getting missed. All of it has a known mechanism, a known frequency, and a known fix, and all of it accumulates while the causes upstream go untreated.

Safety risks become *safety events*, and safety events degrade the quality scores an organization gets judged against. In 2001, the

Institute of Medicine described the gap between care quality that depends on individual heroics and quality that a designed system delivers reliably (IOM, 2001). Where the design is missing, quality becomes a matter of who happens to be working that day. More than two decades later, that is still where much of American medicine sits.

Quality drift becomes *financial loss.* Incomplete documentation produces the wrong billing code and understates how sick the patients were. Quality-linked payments turn negative. Interoperability and information-blocking penalties land on top of both. All of it originates in informatics, and all of it can be treated there. Chapter 3 walks the dollars.

The last link that never appears on a quarterly report is *unwarranted variation*, meaning differences in how patients get treated that have nothing to do with medical evidence. The Dartmouth Atlas has documented it for decades, across regions, across organizations, and between clinicians working down the hall from each other. It tracks to local habit far more closely than it tracks to science (Wennberg, 2010). This is what an organization ends up with when it stops designing care and starts reflecting whatever the software allows people to do.

The chain runs the same way every time. Build decisions become cognitive load. Cognitive load becomes safety risk. Safety risk becomes quality drift. Quality drift becomes financial loss. And, financial loss and quality drift together produce the variation that, over years, becomes the organization's identity.

THE MONEY IS AT THE END OF THAT SEQUENCE.

That is the reason this disease has gone untreated for so long, and it is worth stating plainly because it governs everything that follows. The cost never appears where the damage originates. A build decision made in March produces a denied claim in November, a resignation in February, and a quality penalty the following autumn. By the time any of it reaches a financial statement, the connection to the afternoon where an analyst activated a new alert is completely invisible.

Your chief financial officer sees the loss and cannot trace it backward. This requires an individual trained in clinical practice, healthcare administration, and the systems at the same time, which is still a rare combination and the subject of Part III.

Every step in that chain can be governed. In organizations without a skilled Diagnostician, almost every step is being missed. (See Figure 1.1.)

THE CHIEF COMPLAINT · CLINICAL PRESENTATION

The Deterioration Cascade

How one ungoverned decision becomes organizational disease.

FIGURE 1.1

Six stages, one cascade — and the treatment that interrupts it.

THE DIAGNOSTICIAN'S HISTORY AND EXAM

I have walked into multiple organizations that were struggling badly, three or more years after installing the best EHR on the market. My first question was always the same. Is there a physician executive in charge of this system and the clinical changes that come with it?

When the answer was no, I could describe the environment before seeing it, and I was rarely wrong.

The EHR had been run as an IT project. The clinical outcome that the last several major builds were supposed to change has never been named and measured. Capacity and throughput decisions were being made in rooms where the underlying data were not trusted, and never were sent to be fixed. A governance committee existed and met monthly, and it had never once stopped a thing. Vendor agreements had been signed without reading the sections that moved clinical risk onto the organization. The clinical content had been left on vendor defaults and never matched to an actual protocol. And, the organization lived in a *data desert*. The notes were useless for clinical care, the finance team did not trust the dashboards, every leader was asking why, and the questions were never answered.

The reverse is just as consistent. When the diagnostic process gets followed, when clinical priorities are named first and mapped to what the technology offers, supported by workflows that were intentionally redesigned, governed by a committee with real clinical authority, and measured against defined outcomes, the change arrives quickly and it arrives everywhere. Override rates drop. Time comes back to clinicians. Documentation gets more complete and more accurate. Safety events fall. Quality measures move. Payment improves. And, clinicians report being happier at work.

None of that happened because the technology changed. In most cases it was the same software, from the same vendor, under the same contract. What changed was the relationship between the technology and the clinical work it was bought to serve, and a role finally accountable for that relationship.

That is the argument of this book reduced to its simplest form. The instrument is already bought. Most organizations are already spending eight figures a year to run it. The missing link costs a fraction of that: a physician trained in informatics, given real authority, adequate time, and a team, whose work is to make the instrument serve the medicine.

I will be more precise than a fraction in Chapter 8, where the cost versus gain gets its full accounting. American healthcare spends roughly $67,000 per physician per year on health information technology. The Diagnostician Function that optimizes what that technology does costs about *$2,840*. So, $0.96 of every $1 goes to the instruments. And, $0.04 goes to the hands, if anything goes at all.

Set against what these systems already cost, it is a rounding error. Set against what it can return, it is almost free.

BEDSIDE QUESTIONS: THE HISTORY AND PHYSICAL OF YOUR OWN ORGANIZATION

Answer these honestly. Each is a focused element of the clinical examination of your own organization.

1. Does your organization require that a specific clinical priority be documented before any technology feature is activated or configured?
2. Who governs your clinical technology decisions? Is it a physician with formal informatics training and explicit authority to stop, modify, or require a deployment?
3. For your three largest technology investments of the last five years, can any person state the clinical outcome each was meant to change, and whether it changed?

4. When did your organization last decline to deploy a technology because the clinical case for it could not be made?
5. Can you trace a direct line from your EHR configuration to your quality metric performance?
6. How much do you lose annually to documentation-driven coding deficiencies, inaccurate risk stratification, and denied claims that were eventually paid? All three figures exist in systems you already own. Have they ever put on the same page?
7. Does your organization track health-IT-related safety events as recommended by the SAFER Guides (ONC, 2025)?
8. Is a clinical informatics leader seated at your strategic planning and quality governance tables, not as an invited guest but as a voting member?

The answers to those questions are the chief complaints of this chapter, presenting in your organization.

The chief complaint has been named. The disease is presenting with symptoms across four domains, no Diagnostician has taken the case, and the deterioration follows a predictable arc: cognitive load, then safety risk, then quality drift, then financial loss, then variation that becomes permanent. None of them will reverse on their own. Let us measure the patient's vital signs.

CHAPTER 2

VITAL SIGNS OF AN OPERATING SYSTEM

THE SYSTEM BEHIND THE NUMBERS

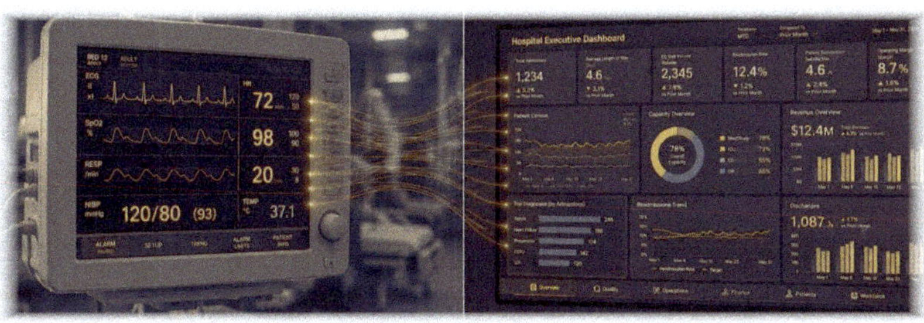

Operational outcome measures are the vital signs of a healthcare organization. They tell you, in real time, whether the patient is compensating, decompensating, or coding.

The emergency department calls bed control at 4:42 in the afternoon to report a 43-bed boarding load and an inbound trauma. Bed control answers that the 9 pending discharges on the medicine floor still don't have orders, transport is 2 hours behind, and environmental services hasn't kept up with turnover since lunch. The nursing supervisor pages the physician on call, who is in a family meeting, then the second call, who is in a procedure, then must decide whether to interrupt the chief medical officer (CMO) at a board meeting. By the time discharge orders flow, an ambulance has diverted to a competitor, two patients have left without being seen, and the trauma is in a hallway bay because the resuscitation room is overflow for a stable but unmoved cardiac admission.

Nothing about this is new. Every physician and hospital in America has lived a version of it. And, none of it is a clinical failure. The medicine is fine. The trauma will be cared for, the boarders will eventually move, the patients will eventually be discharged. What failed is the *operating system* of the hospital: the choreography of beds, orders, transports, and decisions on which the clinical work depends.

In medicine, we have a name for the things we measure to assess whether the operating system is working. We call them *vital signs*.

The pulse. The blood pressure. The respiratory rate. The oxygen saturation. The temperature. They do not, by themselves, make a diagnosis. They tell us how the body is coping with whatever the underlying illness is. A patient with normal vitals may still be very ill, but the body is compensating. A patient whose vitals are abnormal is no longer compensating, and the clock is running.

A healthcare organization has vital signs of its own. Length of stay. OR turnover time. ED arrival-to-provider time. Discharge-by-noon rate. OR first-case on-time start. Outpatient cycle time. Documentation completion time. Inbox turnaround. Revenue-cycle days. Staff vacancy rates. They are all continuous samples of how the organization is coping with the work it has been asked to do. Alone, they are as non-specific as a pulse. Read alongside the exam and the workup, they become diagnostic. And, a drifting number predicts decompensation in systems far removed from the one where it's measured.

Every intern learns two rules about vital signs, usually in the first month of residency, and usually the hard way.

The first rule is: *never ignore a vital sign*. The abnormal value that was noticed, charted, and not acted upon is a fixture of many morbidity and mortality conferences across the country.

The second rule is: *never treat one*. The fever is not the disease. The blood pressure is not the disease. The intern who medicates a number without asking why the number is abnormal, in this patient, at this hour, will eventually harm a patient. *Treat the patient, not the number* is among the oldest aphorisms in medicine, and it has earned its age.

Most healthcare executive leadership has not been taught these clinical rules. Organizations ignore their operational vital signs for quarters at a time, until a board member, a payer, or a ratings agency forces the issue. Then they swing to the opposite error and treat the number itself, with targets, incentives, and dashboards, while the underlying disease progresses untouched.

Both vital sign errors are common. The second one is dangerous. Misusing these measurements is not merely an administrative problem. It creates an iatrogenic hazard, and it has harmed the patient.

WHAT OPERATIONAL VITAL SIGNS TELL YOU

Another clinical pearl, drilled into every internist during training, is: *vital signs are deceptive when read in isolation*. An elevated heart rate of 110 is not a diagnosis. It can represent anxiety, early sepsis, silent hemorrhage, hyperthyroidism, or the first hour of a myocardial infarction. The rate is better interpreted in context. If it stays elevated despite treatment, it is telling you something is being missed.

Operational vital signs work the same way. A length of stay creeping from 4.6 days to 5.1 days over 6 months is not a diagnosis either. It can be acuity drift, discharge-planning failure, documentation lag, a payer-mix shift that changed the case mix, a new EHR build that buried the discharge worklist behind two extra clicks, or a consequence of slower pacing after a beloved hospitalist leaves and a less efficient locum fills the gap. The number alone does not tell you which. Its persistence in the face of effort tells you that something the leadership has not yet identified is preventing the patient from getting better.

A VITAL SIGN IS A SCREAM, NOT A FULL SENTENCE. IT TELLS YOU something needs your attention. It never tells you why. Almost every failed operational initiative and metric-driven project collapse that I have witnessed, began with someone mistaking a scream for a sentence.

TRACING THE VITAL SIGNS

Again, a vital sign is a measurement, not a diagnosis. Nothing in a hospital is sick with a length of stay. You cannot treat a number. You can only treat the system that produces it.

When an internist sees an abnormal vital sign, they do not diagnose the number. They ask which organ systems could generate that finding and proceed to a *review of systems*: cardiac, respiratory, renal, neurologic. The vital signs announce that something is wrong. The review of systems organizes the search for *"where?"*

The operating systems of a healthcare organization work the same way, and there are five of them. *Flow*, meaning throughput and capacity, is the circulation. The *workforce* is the organ reserve, the regenerating capacity everything else draws on. *Communication and coordination* are the nervous system, the signaling between organs. *Documentation* is the work of breathing, necessary, and pathologic the moment it consumes more energy than it delivers. *Operational data integrity* is the brain, the structure on which every other reading depends. Each of them has its own pathophysiology, its own evidence base, and its own modes of failure. These systems improve under clinical informatics leadership and deteriorate in its absence.

Metrics are findings. Systems are where the disease lives.

Throughput and Capacity

Throughput is the rate at which patients move through the care continuum: emergency arrival to bed assignment, OR arrival to incision, clinic check-in to clinician contact, acute episode to discharge. *Capacity* is the size of the corresponding pipes: staffed beds, available ORs, open clinic slots, and qualified personnel.

Throughput failures are among the most studied phenomena in operations research and among the most poorly addressed in healthcare. ED boarding is the clearest example. A patient has been admitted to the hospital. The decision is already made, the plan is already written down, but the bed does not exist. So, the patient stays in the emergency department, in a hallway or a treatment bay, in a unit built to triage and stabilize rather than to care for patients over days. Mortality rises. Length of stay rises. Medication errors rise. Patient experience falls. And, the impact compounds the longer the boarding persists (Rabin et al., 2012; Singer et al., 2011; Asplin et al., 2003; Pines et al., 2011; Morley et al., 2018).

OR turnover delay is not merely an inefficiency. It cascades into late-day cancellations, surgeon dissatisfaction, and revenue loss easily reaching the millions of dollars per OR per year.

In every organization I examine for a throughput problem, I begin in the same place. Not with the dashboard.

With a stopwatch and a pair of sneakers. I walk the path the patient walks. I time the order from the moment the decision is made to the moment the receiving unit acknowledges it. I shadow the bed assignment process. I sit down with the environmental services dispatcher. I observe the discharge planner's actual workflow, not the one in the policy manual. I look at where the EHR helps and where it gets in the way.

By the time I finish, I usually know what the throughput problem is, and I have not opened the dashboard yet. I have also discovered that the staff who do the work can nearly always name the problems before they are formally measured. The measurement simply supplies the magnitude.

The dashboard is necessary. It tells me whether my hypothesis holds at scale. The numbers themselves are rarely sufficient, because they reflect only what was instrumented, and the failure is almost always in something that wasn't. The discharge that did not happen because the consult never returned a recommendation. The admission that bottlenecked because the ED order set has a critical field left blank. The transport delay invisible to the EHR because transport runs on a separate paging system poorly linked to the patient record.

A data analyst cannot see any of that from the analyst's chair. A lead physician walking the same wards as the clinicians and patients often sees it in the first hour.

Workforce and Capacity Mismatches

Capacity is not just available beds. Capacity is also the people who make those beds useful. The workforce that staffs the modern American hospital is the most strained it has been in decades.

The numbers tell part of the story. RN turnover peaked above 27% during the pandemic and settled at 17.6% in 2025, with an average replacement cost above $60,090 per nurse. Travel staffing has become a structural feature of the labor market rather than a transitional one (NSI, 2026). The Association of American Medical Colleges (AAMC) projects physician shortages of up to 86,000 by 2036, with primary care and the surgical subspecialties bearing the largest deficits (AAMC, 2024). Medical assistant, respiratory therapist, pharmacy technician, and environmental services turnover follow similar arcs.

Technology is implicated in both directions. Well-designed clinical workflows, with the right team-based contributions, and the right information available at the right moment, can materially reduce documentation burden, retention loss, and the toll on morale that has come to define the American clinical workforce (Rotenstein et al., 2023; Sinsky et al., 2016).

Poorly designed workflows do the opposite. Melnick and colleagues demonstrated a dose-response relationship between EHR usability and burnout at the population level (Melnick et al., 2020). This finding has unusual clarity in modern occupational medicine. In clinical pharmacology, a dose-response curve like this prompts a dose reduction or a switch in agent. In healthcare, we order a refill.

Turnover is the vital sign with the clearest price attached. NSI's survey of 527 hospitals puts a single percentage point of registered nurse turnover at $294,976 a year for the average hospital, and the average hospital's total annual turnover cost at $5.19 million. The gap between what a travel nurse costs and what an employed one costs runs around $66,081 per position per year.

We will return to this in Chapter 8, to look for the opportunity to intervene.

YOUR WORKFORCE STRATEGY IS A TECHNOLOGY STRATEGY, AND YOUR *technology strategy is a workforce strategy.*

The chief human resources officer (CHRO) cannot fix burnout alone, because half of it is being driven by the EHR. The chief information officer (CIO) cannot fix it alone, because half of it is being driven by clinical team design and staffing models. Moving it takes an executive-physician with visibility across both.

Communication and Coordination

Information must flow in healthcare. Across shifts, services, settings, and across the artificial seam between inpatient and outpatient environments that almost no organization has fully bridged. When it does not, the consequences are predictable, well documented, and directly attributable to design failures the Diagnostician is positioned to address.

Communication failure has been among the most frequently coded contributing factors in Joint Commission sentinel-event reviews for two decades, and its rank has not meaningfully improved across that period (TJC, 2017). *Hospital discharge* is the most fragile transition of them all. The classic Kripalani study found that *direct conversations* between hospital and primary-care physicians at discharge occurred in 3% to 20% of cases, and that the *written discharge summaries* were available less than 33% of the time at the first post-discharge visit (Kripalani et al., 2007). Subsequent work has improved the margins a little, without resolving the structural issues. Formalized handoff systems, like the I-PASS protocol, have cut medical errors during inpatient transitions by 23% where they are implemented with discipline (Starmer et al., 2014).

Vendors increasingly sell automated messaging and alerts as an answer. The literature is consistent on this: *communication problems are not solved by technology alone, but they cannot be solved at all without it.*

The Health Level Seven (HL7) FHIR standard and the SMART-on-FHIR platform built on top of it have finally made it possible to push the right information to the right person at the right moment without building a custom interface engine. A capability does nothing on its own. The organization must eventually identify which information matters, in what format, at which care transition, which clinicians will read it, and where the path can lead to inattention or misreading (Mandel et al., 2016).

Closed-loop communication requires messages that are meaningful, timely, and correctly prioritized, sent by a route the recipient reliably reads, with the receipt confirmed.

Too many loops stay open.

The technology is only the medium. This is a human factors issue. One person writes the message, and some other person has to act on it, and nearly everything that fails between those two points fails by bad design before it fails by bad behavior. Designing it is informatics work.

Documentation Burden as Operational Drag

Documentation has become, for many clinicians, the largest single component of the working day. The 2016 Sinsky time-motion study famously demonstrated that for every 1 hour of direct patient care, physicians spent nearly 2 additional hours on EHR and desk work, with 1 to 2 hours more after the clinic closed (Sinsky et al., 2016), which some refer to as "*Pajama Time.*" Rotenstein's 2023 longitudinal follow-up at two academic medical centers confirmed the burden has not abated and may have even worsened. Primary care physicians spent a median of 36.2 minutes on the EHR per visit, and the variation across clinics tracked primarily to system-level factors like how teams contribute to order entry, support for medication refills, and clinic staffing ratios. Almost all of those sit within the clinical leadership's power to change (Rotenstein et al., 2023).

The operational consequence is brutal. Every minute in front of the EHR is a minute not spent at the bedside, not spent in clinic, not spent recovering before the next encounter.

The organization pays for clinician time and gets documentation time. It pays for documentation time and gets a record the next clinician finds unreadable.

This is not a minor operational issue. More than any other, it determines whether the workforce stays, and whether the organization can keep doing the volume its community needs.

It is also the operational issue most directly within the Diagnostician's reach. Note redesign, template simplification, order-set rationalization, inbox management, and the deployment of voice and ambient documentation under rigorous evaluation are all clinical informatics interventions, and properly governed they return measurable time to the clinician. In recent ambient AI work, reductions in self-reported burnout are encouraging (Tierney et al., 2025a).

Operational Data Integrity

Operational dashboards lie when they are built carelessly. Most are constructed to only track what the EHR can easily measure, rather than what the organization's decision-makers really need. The result is a competent leadership team making confident decisions on data which do not reflect reality.

In many organizations, some of the most consequential operational decisions are made using numbers that have never been audited. Throughput is reported off audit logs that capture order placement rather than order completion. Length of stay is reported off discharge-disposition fields populated hours or days after the patient has left. OR utilization is reported off room-turnover times that begin and end at administratively convenient moments.

None of this is intentional.

It is the natural drift of measurements away from their reality whenever the people who build the measures are not the people who do the work. It is also one of the most reliable diagnostic findings of an organization without a Diagnostician. (See Figure 2.1.)

TAKING THE HISTORY · OPERATIONAL VITAL SIGNS

The Five Systems Behind the Vital Signs

A metric is a finding. The system is where the disease lives.

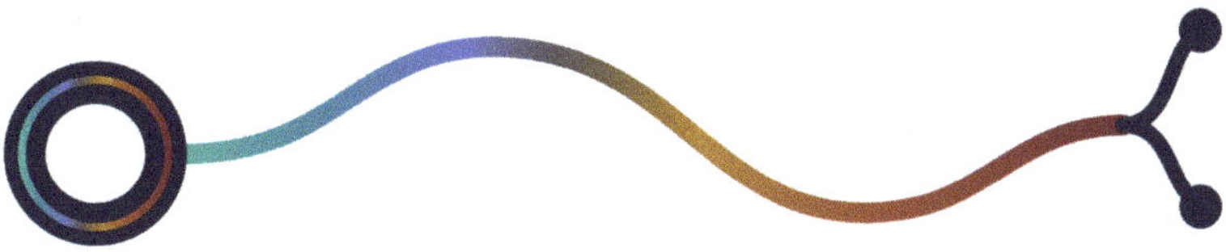

#	System	Metrics
1	Throughput & Capacity	LOS · ED arrival-to-provider · OR turnover · discharge-by-noon
2	Workforce	Vacancy rate · turnover rate
3	Communication & Coordination	Page response time · open consults · transport delays
4	Documentation	Note-completion time · inbox turnaround
5	Operational Data Integrity	Data validation audits · source-of-truth checks

Two rules every intern learns the hard way: never ignore a vital sign, and never treat one. Treat the patient, not the number.

FIGURE 2.1

The operating system has five systems; every deranged metric traces to one.

THE DISCIPLINE OF INTERPRETATION

None of the vital signs above will be new to an executive reading this chapter. They are already on a dashboard somewhere in your organization, refreshed nightly, reviewed monthly, presented quarterly. The failure is not only in the data collection process. It is often also in the interpretation, which has rules that parallel clinical diagnostic skills. They all translate directly from the bedside.

Read the entire panel, not just one number.

A heart rate of 120 means one thing when the blood pressure is normal and another thing entirely when the pressure is falling: the first is an isolated finding, the second is the trajectory.

The same is true upstairs. A length of stay drifting upward at the same time RN vacancies are rising and claim denials are climbing is not three problems for three different vice presidents to address. Often, it is one disease affecting all three findings. For example, a short-staffed and exhausted floor discharges late; an exhausted workforce also documents thinly; thin documentation fails payer review; failed review reduces the reimbursement. The organizational chart distributes those findings to three different owners. The disease does not respect that distribution.

THIS IS WHY THE SAME ORGANIZATION FUNDS THREE SIMULTANEOUS improvement projects, one per silo, and treats one illness three times badly.

Trending beats threshold.

One of the most dangerous values on any vital signs panel is the normal one moving subtly and steadily in the wrong direction. A systolic blood pressure of 110 is reassuring unless it was 140 an hour ago, in which case it may be a hemorrhage announcing itself while still inside the reference range. Operational dashboards, with their red-yellow-green thresholds, are engineered to miss exactly this kind of finding. The green box rewards the snapshot and conceals the slope.

Current dashboard thresholds deserve more suspicion than they usually get, and I have rarely heard this discussed openly. Most healthcare thresholds are based upon benchmarks, and a benchmark tells you *where the herd is* rather than *whether the herd is healthy*. U.S. hospitals have a median margin of just 2.9% (Kaufman Hall, 2026), while more than 40% of their physicians meet burnout criteria (AMA, 2024).

That is not a herd whose 50th percentile deserves to be a performance target.

Every abnormal value earns a differential, not a verdict.

When RN turnover rises, the list of possible causes is quite long: low market compensation, high scheduling burden, low span of control, safety concerns, high documentation load, the quality of their first-line manager, or the absence of a career pathway. Each is a different disease, with a different treatment, at a different price tag.

The list of possible causes of an abnormal finding often collapses to whatever the accountable executive understands and has the power to treat. Human resources diagnoses compensation, because compensation is the lever HR holds. The CIO diagnoses technology usage. Finance diagnoses the staffing model. Those diagnoses are an artifact of the organizational chart rather than a finding of the examination. They all produce a treatment, an expense, and no improvement.

No internist would let the differential diagnosis for chest pain be narrowed by which specialty was easiest to reach on the phone. Healthcare organizations inadvertently make decisions exactly that way.

Abnormal vital signs are often linked.

An easy example is the patient who is losing blood. One cause, the blood loss, makes the pulse and respiratory rate high, and the blood pressure and urine output low. Looking for four different causes of those four different values takes you off course. Stopping the blood loss fixes four abnormalities at once. Treating any of those four by itself will result in a bad outcome.

Hospital vital signs, in the same way, often share one underlying physiology. This means that treating one is frequently the cause of another's abnormality. Pushing length of stay down by quickening discharges will often show up as an increased 30-day readmission rate within the quarter. The patients discharged before they were ready come back through the front door. Increasing the volume of filled patient appointments without examining your capacity will show up as an increase in the left-without-being-seen rate. More volume stresses capacity.

The panel is a closed system. A leader who improves one number without knowing where the displaced pathology went has not treated the patient, they just moved the symptom to another patient's chart.

Re-examine after every dose.

No physician orders an intervention without ordering the re-examination: the repeat lactate, the post-transfusion hemoglobin, the vitals every 4 hours. The intervention is based upon a probable hypothesis. The re-examination is the test of that hypothesis.

Operational interventions in most health systems are launched, celebrated, and closed when the go-live completes rather than when the vital signs move. "Implemented" gets recorded as the outcome, and the measurement that motivated the whole effort is never consulted again.

An organization that does not re-examine after treating cannot learn which of its treatments work. That is how a health system arrives, a decade later, holding a portfolio of annual initiatives and a set of vital signs that have not improved in any of those years.

THE DANGER OF TREATING A NUMBER

We have briefly discussed a common rule in clinical vital sign interpretation: *treat the patient, not the number*. There are many reasons why isolated numbers mislead. The monitor may not be calibrated, the data may not be correctly collected, or they may feed into a decision-making bias.

The executive equivalent can also occur. The evidence on what happens when organizations treat their numbers is extensive, consistent, and largely ignored by an industry that needs it.

The underlying principle was described decades ago. Campbell, studying public-sector measurement, formulated what is now called *Campbell's law*: the more any quantitative indicator is used for decision-making, the more subject it becomes to corruption pressures, and the more apt it is to distort and corrupt the processes it was intended to monitor (Campbell, 1979). Economists know the same finding as *Goodhart's law*, usually paraphrased as: when a measure becomes a target, it ceases to be a good measure (Strathern, 1997). This is among the most replicated findings in the study of organizations (Muller, 2018), and healthcare has supplied some of its most painful confirmations.

The natural experiments are on the record. The English National Health Service attached aggressive targets and executive consequences to its four-hour emergency department wait times. Reported performance improved. And, so did the artifacts. Documented wait times clustered suspiciously just inside the threshold. Ambulances were held outside departments so the clock would not start. Patients were moved to "assessment units" minutes before breaching the threshold (Bevan and Hood, 2006).

The number was treated. Whether the patients were is a separate question, and that is precisely the problem.

At Mid Staffordshire, the public inquiry concluded that a hospital board focused on financial and access targets, while presiding over wards where the quality of care had collapsed without notice. The numbers the board reviewed were reassuring nearly until the end (Francis, 2013). In the American version, the Veterans Health Administration tied a 14-day scheduling standard to executive performance pay and produced long waiting lists engineered to satisfy the metric rather than to improve actual 14-day access (VA Office of Inspector General, 2014).

All three are sometimes remembered as failures of character. All three are better understood as failures of method. A vital sign converted into a target and never re-examined rigorously. The character failed downstream of the faulty design.

The subtler American versions are running now. The Hospital Readmissions Reduction Program penalized a vital sign, the 30-day readmission rate, and the vital sign improved. Whether the patients improved remains seriously contested. Analyses of Medicare data reported an associated rise in post-discharge mortality for heart failure and pneumonia in the penalty era (Wadhera et al., 2018). That finding is debated, and these pages are not the place to settle it. The hypothesis it raises is the point, and it should keep every executive who has ever set a readmission target awake at night.

A hospital can lower its readmission rate either by improving discharge care, or by becoming more reluctant to readmit the patients who need to come back. The metric cannot tell the difference. Only an examination of the patient can.

The sepsis bundle tells a similar story in miniature. It was built from genuinely lifesaving elements. Once enforced on a clock and tied to public reporting, it drew formal objection from the Infectious Diseases Society of America on the grounds that it pressures physicians toward broad-spectrum antibiotics for patients who do not need them (Rhee et al., 2021). Compliance rises. Stewardship erodes. The number is treated.

The recent focus on the "expected" mortality rate addresses the same problem in a different way. An organization can improve its observed-to-expected mortality ratio without preventing a single death. Document the severity of illness more completely, and the expected deaths in the denominator rise. The ratio falls. The board slide turns green. The effort is legitimate, because incomplete documentation genuinely understates how sick the patients were, and correcting it corrects the record and the risk. But an executive team can mistake denominator repair for clinical improvement and never look for the deaths it could prevent.

Clinicians and frontline staff do not game metrics because they lack integrity. They game metrics because the organization has made the metric the priority. Attaching the number to a bonus and a strategic initiative has communicated that the number is the thing to be treated. Frontline staff are rational. They treat what leadership demonstrably cares about.

An executive who attaches meaning to a vital sign and begins to address the numbers without a clear diagnosis is prescribing without examining the patient. Medicine has a name for that, and a licensure system built to prevent it.

DIAGNOSIS OF THE VITAL SIGN READER

So far, this chapter has treated healthcare vital sign issues as a design problem, and much of it is. But every vital sign panel ends at a reader. Years of sitting across the table from the readers have taught me that they come in different forms, distinguishable by their response to an abnormal finding.

An underprepared reader is often not aware of the abnormal value or what it implies. If tasked, they struggle to form a differential diagnosis, because they do not know what questions the number answers. Their signature response is broad *empiric therapy*, aiming at everything in the hopes of hitting something: the staff reorganization, the consultant engagement, the system-wide initiative, all broad-spectrum treatment for an undiagnosed disease. They are not a bad actor. They are practicing above their license, without ever having been told.

A complacent reader receives the abnormal value without curiosity or concern. They do not focus on or question the number because they do not own the metric as theirs. The dashboard is reviewed because the agenda says the dashboard is to be reviewed, and the meeting always ends on time. This is the organizational equivalent of the vital sign noticed, charted, and not acted upon, and it produces the same morbidity and mortality finding delayed.

The competent and motivated reader treats the abnormal value the way a doctor treats a chief complaint. They order the workup. They walk the floor themselves or send a colleague who knows what to look for. The distinguishing sign on examination is that they ask for help in public, the way a good internist calls a consult. *Competence at this level is not measured by knowing every answer but by understanding the question well enough to engage whoever can answer it.*

Finally, there is a fourth phenotype, which the measurement literature rarely names and every experienced executive has met.

The reader with conflicting incentives. The fourth phenotype requires no malice to explain, only an ordinary human incentive under extraordinary pressure. The reader with conflicting incentives does not misread the vital signs at all, they read them accurately and understand exactly what the numbers mean. What matters to them is not what the data reveal but what they imply about their own tenure. The incentive quietly inverts. The goal stops being to fix what the data show and becomes to manage what the data are allowed to show. They already know, privately, that something serious is wrong on their watch. What they cannot tolerate is the public record of it.

The findings on examination are consistent, though easy to miss. Metric definitions left just confusing enough to be contestable whenever they turn unfavorable. Analytics leadership that turns over with curious regularity. The messenger getting reassigned more often than the message getting addressed. Board packets that read better each quarter while the wards tell a very different story.

The first two phenotypes are easily treatable. *Lack of knowledge* responds to education. *Complacency* responds to governance that makes the workup mandatory. Both improve further with access to a Diagnostician at the table whose consult the executive learns to call.

The fourth is not treatable, because every instrument that could detect it reports to them. It requires a board-level diagnosis, and the board cannot make it from a curated packet.

The prescription for all follows from everything this book has argued so far. An abnormal value should trigger a workup, not an immediate conclusion. Perform no intervention without an identified diagnosis, because aiming at nothing is a way to cause inadvertent harm. Pair every metric with balancing metrics reviewed by the teams that "own" them: readmissions with post-discharge mortality, length of stay with readmissions, increased clinic volume with left-without-being-seen, exactly as a physician pairs the diuretic with tomorrow's creatinine.

And, keep a physician-leader with diagnostic training and operational authority between the dashboard and the decisions. The difference between a vital sign used well and a vital sign that harms is not the number, it is the discipline of the person reading it.

WHAT NORMAL VITAL SIGNS LOOK LIKE

When these vital signs are read together as a panel, trended rather than as thresholds, worked up rather than judged, and re-examined after every intervention, every domain on the operational dashboard begins to move. Boarding falls because the upstream and downstream blockers, finally visible, are addressed in concert. Documentation burden falls because templates are redesigned around how clinicians think rather than what the vendor demonstrated at the conference. Communication improves because handoffs are designed by the people doing them, supported by the technology rather than constrained by it. The workforce stops leaving as fast.

One more finding belongs in that picture. In a healthy organization, the people who do the work trust the numbers. The charge nurse recognizes the dashboard as a description of the shift they worked. The metric review opens with "What is this number telling us?" rather than "Who owns this number?" No vital sign provides an incentive to favor it in the differential.

This is the predictable result of doing the workup before writing the prescription, and of remembering, at enterprise scale, the two rules every intern learns in the first month.

BEDSIDE QUESTIONS: THE HISTORY AND PHYSICAL OF YOUR OPERATIONS

1. Which of these five systems is your organization currently most worried about, and what is the trend in the vital signs it generates over the last 6 months?
2. Who governs the definitions of those metrics? Are the definitions stable across departments, or is each leader looking at a slightly different number?
3. When was the last time a senior leader walked the workflow that produced one of those numbers, with a stopwatch and without the dashboard?
4. Which of your operational metrics are tied to executive or departmental incentives, and for which of them has a documented differential diagnosis ever been completed?
5. Has any metric improved in the past year while the frontline experience of the same workflow worsened? How, specifically, would your organization detect that?
6. Have your operational dashboards ever been audited

against the actual workflow they purport to measure in the last 12 months?

7. Is your Diagnostician (CMIO or equivalent) seated at the operations meeting as a peer, or invited as a guest?

An organization that cannot answer these questions is not reading and responding to its own vital signs.

THE VITAL SIGNS ARE ABNORMAL IN RECOGNIZABLE PATTERNS: boarding, turnover delay, documentation debt, attrition. The Diagnostician knows how to read them and what their deterioration predicts. The laboratory is the next stop. The numbers there will tell you what the data are hiding.

PART II

THE WORKUP: LABORATORY, IMAGING, AND CONSULTS

The history, physical and vital signs are suggesting a syndrome. The next step is the workup. Part II walks through the diagnostic instruments: the laboratory markers of margin, documentation, and reimbursement; the EHR as an imaging modality; the data quality on which every dashboard depends; the iatrogenic-harm review every sick organization eventually requires; and the specialty team assembling the picture. The findings will either confirm the diagnostic impression or suggest the alternatives the history and physical alone could not rule out.

CHAPTER 3

LABORATORY MARKERS OF SURVIVAL

WHAT IS THE NUMBER

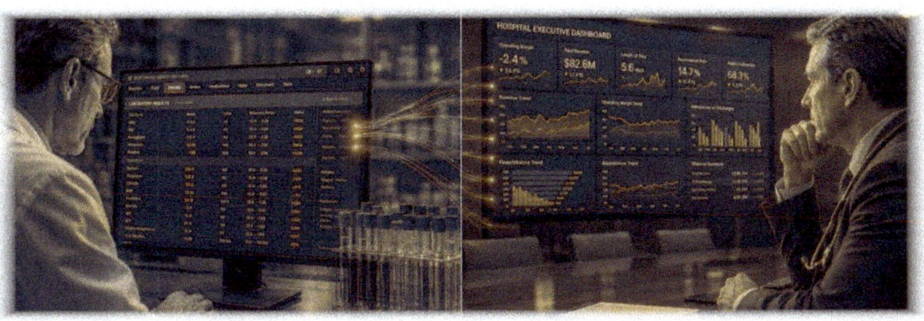

A healthcare organization's financial lab values are not a separate concern from its clinical and operational reality. They are the downstream consequence of clinical and operational reality, captured imperfectly through the lens of documentation and coding.

The morning lactate lab value comes back at 4.6 mmol/L, more than twice what it should be, and to the resident covering overnight, it could only mean sepsis. The patient is confused, breathing fast, and by every protocol on the unit an elevated lactate in a confused patient triggers the same response, fluids running wide open, blood cultures drawn, broad-spectrum antibiotics hung within the hour. The number told a story, and that story was infection.

Except a lactate does not tell a story. It confirms that something in the body is under enough stress to burn fuel the wrong way, and it says nothing about what that something is. A seizure burns lactate the same way an infection does, and when the nurse mentions, in passing, that the patient's roommate heard a strange rhythmic sound from the bathroom an hour before the confusion started, the picture changes entirely. No fever. No source. A tongue bitten on the side. The antibiotics meant to treat the sepsis were never going to treat a seizure that had not been witnessed or questioned.

The lactate had been accurate the whole time. It never lied or knew what caused it. The treatment addressed a disease the patient did not have, while the one they did have went unexamined for another hour. A number that can mean many things is not a diagnosis. It is a reason to keep asking questions.

A LAB VALUE IS A NUMBER

A patient with severe anemia does not need a transfusion because the hemoglobin is low. The patient needs a transfusion because the hemoglobin is low *and* the patient is symptomatic, hemodynamically unstable, and likely still bleeding. The number is just a measurement. The decision requires synthesis.

Like an isolated vital sign, the treatment of an isolated lab value will eventually harm a patient.

Healthcare leaders are routinely encouraged to treat the lab value. The board sees a margin number. The bond rating depends on a margin number. The compensation committee benchmarks against a margin number. So, the boardroom discussion centers on margin in isolation, as though margin were the disease.

It is not. Margin is a downstream measurement of how the clinical and operational reality of the organization is being captured, coded, billed, and paid for. Treating margin without understanding the upstream pathology is the financial equivalent of transfusing the anemic patient without finding out if they are bleeding.

The financial lab values are diagnostic only when the Diagnostician knows where the bleeding is coming from. This is the first study of the workup.

WHAT THE CURRENT LAB VALUES MEAN

The laboratory profile of a sick American hospital, in 2026, has a recognizable pattern. Median operating margins are still recovering from the post-pandemic trough, when more than half of U.S. hospitals reported negative margins (AHA, 2022), and even now sit at just 2.9% (Kaufman Hall, 2026). Bad-debt and uncompensated-care percentages have crept up. Days-in-accounts-receivable are inching higher. Denials, in many systems, have outpaced inflation (AHA, 2026). Capital constraints have produced deferred maintenance, slowed recruitment, postponed expansion, and in a growing number of communities, closed services those communities depend on.

The lab values themselves are not in dispute. What is in dispute is the *diagnosis*.

The boardroom diagnosis is usually some version of "we are not running tight enough." The chief financial officer (CFO) is asked to find more savings. Service lines are asked to improve their margin. Recruitment is paused. Capital is throttled. None of that is wrong, exactly. None of it touches the upstream pathology that produced the lab values in the first place.

Before naming that pathology, there is a prior question the boardroom almost never asks. *Where does all the money go*? The answer is not where most people assume, and getting it wrong makes every decision downstream of it harder.

UNDERSTANDING HEALTHCARE'S COST

Cost discussions in American healthcare reflexively blame the doctor and the hospital. The arithmetic does not support the blame, but the reason it does not is subtle. The actual drivers are rarely named in boardrooms because they sit on different lines of the income statement.

The Conventional Story

The conventional story runs roughly as follows. American physicians are paid more than physicians in peer countries. American hospitals operate at higher margins. American care is delivered with less efficiency, more variation, and more administrative overhead. Therefore, the cost problem is a doctor-and-hospital problem, and the remedy is to constrain physician compensation, hospital pricing, and the operational decisions that produce them.

Parts of this story are true. Physicians and hospitals are not innocent contributors to the country's $5.3 trillion annual health expenditure (CMS, 2025c). Compensation in some specialties has outpaced inflation. Hospital charge-master pricing is, in many institutions, indefensible. Variation in clinical practice is real and consequential. None of that should be denied. Good clinical informatics governance lends itself to the discipline of cost containment.

The conventional story is also built on a category error, and the error is easy to make because the national arguments invite it.

WHERE HEALTHCARE'S MONEY GOES

In 2024, the United States spent $5.3 trillion on healthcare, 18% of gross domestic product (CMS, 2025c). The two largest categories are exactly where the conventional story looks.

Hospital care consumed 31% of the total, roughly $1.6 trillion. Physician and clinical services consumed 21%, roughly $1.1 trillion. Together they account for more than half of national health expenditure, and a reader who stops there will conclude that the conventional story is obviously correct (CMS, 2025c; KFF, 2025; Gunja et al., 2023).

Go one level down and it stops being as obvious.

Physician Compensation

Physician compensation, in aggregate, across hospital, ambulatory, and academic settings, accounts for somewhere between 8% and 10% of national health expenditure (Gottlieb et al., 2023; Keehan et al., 2024). That is the entirety of what American physicians take home, and the number absorbing most of the political and journalistic attention paid to healthcare cost.

The other 40-odd points inside those two lines are not physician income. They are purchased *administrative inputs* and apparatus, riding on the clinical activity and booked alongside. Put another way, "physician spending" in these conversations is really three things added together: physician compensation, administrative overhead, and services quietly booked to the same cost center.

Uwe Reinhardt made the identical arithmetic explicit two decades earlier, in a letter correcting a Times editorial that had proposed balancing the deficit by cutting what doctors are paid. The physician and clinical services line, he wrote, is only about a fifth of national health spending, and close to half of that fifth never reaches a physician, absorbed first by practice expense and malpractice premiums (Reinhardt, 2007; Reinhardt, 2012). The equation has not changed since. Only the industries hiding inside it have multiplied.

Drugs

Retail prescription drugs are reported at 9% of national health expenditure, about $467 billion in 2024 (CMS, 2025b). That figure systematically understates the pharmaceutical share, because drugs given in inpatient settings, physician offices, and infusion centers are booked under hospital or physician services rather than under the prescription-drug line. Reconstruct the full category, and provider-administered and non-retail drugs add roughly 4 to 5 more points, bringing total drug spending to approximately 13% to 14% of national health expenditure at net, post-rebate prices (Brookings Hutchins Center, 2022; Peterson-KFF, 2025). The proportion is rising quickly as biologics and specialty therapeutics displace traditional small-molecule drugs.

Devices, Supplies, and Durable Medical Equipment

As the national accounts report them, durable and non-durable medical products together account for a low single-digit share. The reported figure understates the real one for the same reason drugs are understated: implants, surgical supplies, and high-acuity disposables are consumed inside hospital and physician encounters and are booked there. Growth has been steady, driven substantially by orthopedic and cardiovascular implants whose marginal clinical benefit, in many indications, is contested.

Administration

This is the category that cannot be added to the others, and it is the reason the arithmetic of healthcare spending confuses people. Administrative cost does not sit in a separate column beside hospital care and physician services. It sits *inside* of them, alongside the payer-side overhead the American financing system requires.

The measurements are not subtle. Woolhandler and colleagues put administration at 31% of United States health expenditure in 1999, against 16.7% in Canada (Woolhandler et al., 2003). Then 18 years later, Himmelstein and colleagues measured it again: $812 billion, or $2,497 per American per year, 34.2% of health expenditure against $551 per capita and 17.0% in Canada (Himmelstein et al., 2020). Sahni, Carrus, and Cutler have since estimated that roughly a quarter-trillion dollars of it could be removed with known simplification measures (Sahni et al., 2021).

Two decades of measurement, and the American number went up.

Healthcare IT

The category this book is most aligned with is smaller than most people assume. Total United States healthcare provider technology budgets run near $69 billion, which against $5.3 trillion is roughly **1.3%** of national health expenditure (Forrester, 2026). Measured at the institution rather than the nation, hospital information-technology expense runs near **2.3%** of total hospital operating expense (Definitive Healthcare, 2024).

Two things follow, and they matter more than the size of the stated number. The *growth rate* is among the fastest of any category in healthcare. And, the true figure is almost certainly higher than either measure captures, because a substantial share of technology cost is never *labeled as technology cost*. These costs are usually bundled into depreciation schedules, vendor service contracts, consulting, cybersecurity premiums, and implementation labor charged to operations. The spend is real but booked somewhere other than where it belongs. Very little of it appears on a line called health IT. (See Figure 3.1.)

NATIONAL HEALTH EXPENDITURE · 2024

The $5.3 Trillion Patient

$5.28T · 17.9% of GDP — where the dollars actually go

Administration *(cross-cutting)* — **34.2%**
Billing, coding, prior authorization — booked inside the categories below

Hospital Care — **31%**
Labor, supplies, equipment

Other — **~30%**
Long-term care, insurance overhead, public health, equipment

Physician & Clinical Services — **21%**
Only 8–10 points is physician take-home pay

Pharmaceuticals — **13–14%**
Retail plus provider-administered drugs

Devices, Supplies & Equipment — **2–4%**
Implants, surgical supplies, disposables

Health IT — **~1.3%**
EHRs, AI, analytics, cybersecurity

THE LEVERAGE

Health IT is the smallest category here — and every other line runs through it.

A NOTE ON CATEGORIES

Administration is cross-cutting, not additive. Shares are independently sourced and won't sum to exactly 100%.

FIGURE 3.1

Healthcare spending by category, 2024, ranked by share of national health expenditure.

THE UNTREATED DIAGNOSIS

In 2019, Shrank, Rogstad, and Parekh published an accounting of *waste* in American healthcare. They reviewed 71 estimates from 54 sources across the 6 waste domains defined by the Institute of Medicine, converted everything to annual national figures, and totaled it. Waste came to between $760 and $935 billion a year, as much as 25% of everything the country spends on health (Shrank et al., 2019).

Then they did something most reviews do not. For each domain, they went looking for interventions that worked to reduce waste, and they priced the achievable savings. Failure of care delivery, failure of care coordination, over-treatment, pricing failure, and fraud and abuse all yielded something. Summed, the recoverable savings came to between $191 billion and $286 billion. Skilled clinical informatics, configuration, and workflows can have an impact on all of these.

One domain contributed nothing to that figure.

Administrative complexity was the largest single source of waste they found, at *$265.6 billion* a year. It was also the only domain for which they could not locate a generalizable study demonstrating how to reduce it. They emphasized that the administrative complexity category was associated with the greatest contribution to waste, yet there were no generalizable studies that had targeted it as a source for waste reduction. Their savings estimate simply excludes it.

In clinical terms, the largest identified pathology in American healthcare has no described treatment in the literature.

That is not because the problem is mysterious. Cutler's proposals for reducing it are concrete and unglamorous: a clearinghouse for bill submission, simplified prior authorization, harmonized quality reporting, and better data interoperability (Cutler, 2020).

A clearinghouse is an information exchange problem. Prior authorization is a workflow and structured-data problem. Harmonized quality reporting is a measure-definition and data-capture problem. Interoperability is a standards problem.

EVERY REMEDY PROPOSED FOR THE SINGLE LARGEST SOURCE OF WASTE IN American healthcare is an informatics problem.

And, yet, in most American healthcare organizations there is no executive-physician in the C-suite empowered to address them.

THE PIGGYBACK INDUSTRIES

Set the categories side by side and a structure appears. Most of what flows through the hospital and physician lines is not physician income. This revenue goes to a set of industries whose business depends on clinical activity happening, and whose pricing power is largely independent of the clinical work that activity is meant to deliver.

Call them the piggyback industries.

Pharmaceutical manufacturers ride on physician prescribing and hospital formulary decisions. Their pricing, particularly for biologics, oncology agents, and newer specialty therapeutics, bears little relationship to their cost of production. They are constrained primarily by what payers will reimburse and what the patent envelope will protect. The physician orders from a formulary they didn't negotiate. The patient never sees a price. Neither of them influences what the medication costs.

Medical-device and supply manufacturers ride on procedural decisions, implant choices, and operating-room workflow. The dynamics are similar: substantial pricing power, weak price competition in many categories, and a sales infrastructure organized to protect both. Here the surgeon chooses from a contract negotiated in a different building, often without knowing what any option costs, and the price gap between two clinically acceptable implants can exceed what the hospital is paid for the case. Guidehouse put the annual supply chain savings opportunity for U.S. hospitals at $25.7 billion (Guidehouse, 2019).

Capital-equipment manufacturers ride on the steady demand for imaging, robotics, monitoring, and diagnostic infrastructure. Pricing is opaque, service contracts are substantial, and replacement cycles drive a meaningful fraction of hospital capital budgets.

The administrative apparatus is the largest piggyback of all, and the strangest, because much of it exists to move money rather than to deliver care. Prior authorization, claims adjudication, denial management, appeals, coding review, and payer negotiation employ enormous numbers of people on both sides of every transaction. None of it touches a patient. All of it is booked inside the hospital and physician lines that the conventional story reads as clinical cost.

Health IT vendors are the newest of them and, measured by growth rate, the most aggressive. EHR licensing, implementation services, optimization consulting, AI add-ons, cybersecurity infrastructure, interoperability services, analytics platforms, patient portals, ambient documentation tools, and the consulting and integration apparatus around each of them now form a permanent operating cost that did not exist a generation ago. Growth is structural. Pricing power is substantial. And, whether each new investment produces measurable clinical or operational value is, as this book keeps finding, rarely even asked.

The piggyback industries are not villains. They produce real value. The pharmaceutical that cures the disease, the implant that restores the joint, the imaging study that finds the cancer, the EHR that supports the encounter, the AI model that catches the deterioration, are each a genuine clinical good.

The point is not that these industries should not exist. The point is that the money attributed to doctors and hospitals mostly passes through doctors and hospitals on its way somewhere else, and an honest cost conversation must start there.

WHY THE MISATTRIBUTION PERSISTS

The misattribution of healthcare spending persists for predictable reasons.

Physician and hospital costs appear on the bill the patient, payer, and media see. Pharmaceutical, device, administrative, and technology costs are embedded in line items few ever see. Political and journalistic attention follows visibility rather than magnitude.

Physician and hospital costs are negotiable in payer-provider contracts. Pharmaceutical and device pricing in the United States is structurally protected from negotiation in ways other developed countries do not match. So, the piggyback industries that cannot be negotiated grow at whatever rate the market sets, while the smaller categories that can be negotiated absorb the policy attention.

Physicians and hospitals are organized into licensing structures, professional bodies, and regulatory regimes that make them legible objects of policy. Pharmaceutical and device manufacturers are organized into trade associations and lobbying apparatus that make them legible objects of influence. The asymmetry is structural, well-funded on both sides, and unlikely to change without political reform.

The technology piggyback is nearly invisible outside of the chief information officer's office. The board sees the EHR contract once every several years. The board does not see the depreciation, the recurring AI subscriptions, the cybersecurity premiums, the analytics-platform fees, the interoperability service charges, or the consulting engagements that surround each go-live. The growth in cost is real, substantial, and spread across enough line items that it never appears as one number requiring defense.

WHAT THIS MEANS FOR THE EXECUTIVE READER

Two implications follow these numbers.

First, when board conversations turn to cost containment, the Diagnostician asks where the cost is occurring. Physician compensation is not a primary source. The purchased inputs and the administrative apparatus riding inside the clinical lines should be examined.

Second, health IT, while the smallest of the line-item categories examined here, is also the one most directly within the executive's reach. Roughly 1% of national health expenditure, 2% to 3% of a hospital's operating expense, growing faster than almost anything else on the budget, scattered across enough line items to obscure an actual number, and governed by no owner in particular.

That combination should be reason enough to govern it carefully.

The larger reason is that the 1% health IT category is understated by more than the topline figure admits. Purchasing the tools is just the front edge of the cost. That is the cost of purchasing and turning them "on." The other costs have become embedded in the infrastructure and operation.

Healthcare technology is not passive infrastructure like a concrete slab under the ER. Unlike physical buildings, digital systems require constant setup, training, maintenance, updates, and upgrades managed by the IT team just to keep them running. That is important, ongoing, and costly work.

Then, organizations must push beyond turning basic functionality "on" to configuring these tools for high performance across all departments. This also comes with a significant cost.

These technologies have become the malleable infrastructure through which all clinical care and the 34% administrative category are executed (Himmelstein et al., 2020). Every prior authorization is submitted through them. Every claim, every denial, every appeal, every quality measure, every coded diagnosis, every record request between two organizations that cannot exchange data cleanly.

ADMINISTRATIVE COMPLEXITY IS NOT ADJACENT TO INFORMATION technology. It runs on and through it.

Configure the instruments badly and the largest source of waste in American healthcare grows. Configure them well, with accountability for the clinical and operational result, and it shrinks. Nothing else on the budget offers that leverage, and no other executive is specifically trained to exercise it.

This is not an argument against technology investment, it is the opposite. Technology is here to stay, and it is creating a deepening dependency. The argument is for governing the technology investment with the rigor this book describes, because more than any other line on the budget, this is the one an executive can either spend well or bleed.

The lab values point at this system, one the conventional story does not usually examine. The Diagnostician's job is to write the treatment plan accordingly.

THE PATHOLOGIES OF MARGIN AND LOSS

Now the pathology itself. In nearly every organization I have examined, margin loss traces to three findings, and all of them sit inside the reach of clinical informatics.

The Critical Lab Value for Healthcare Documentation

Reimbursement, in the modern American healthcare environment, is not a function of what was done.

Reimbursement is a function of what was documented.

A patient with severe sepsis whose physician writes "infection" in the assessment is, for billing purposes, a patient with a generic infection, earning a much lower-paying diagnosis-related group (DRG) and risk profile than the sepsis would warrant. A patient with stage IV chronic kidney disease whose primary-care physician writes "CKD" without staging looks, for risk-adjustment purposes, like a patient with mild kidney disease. The technical care delivered may be excellent. The financial consequence will not be.

The direction of the error matters, and it runs one way. American healthcare systematically under-documents the severity of what it treats, and the payer keeps the difference. Clinical documentation integrity programs (CDI), when integrated with the EHR and governed by clinical leadership, recover substantial revenue in study after study, with mature programs documenting annual recoveries between $1.5 million and $10 million per facility. The range varies primarily with how seriously the program is built into the clinical workflow (ACDIS, 2024; Dougherty et al., 2013).

The mechanism is not aggressive billing, but accurate billing, and the matching of the documentation specimen to the clinical reality.

The problem is that documentation, in most organizations, has been handed to the EHR as though the EHR could do it alone. Templates are designed for billing capture rather than clinical reasoning. Physicians, facing the resulting cognitive load, default to copy-forward and macro-driven shortcuts. The note grows longer and says less, and it becomes a steadily poorer record of the care delivered. The financial lab value drops. The boardroom cannot name the pathology, because they do not read clinical notes.

The Diagnostician has read the notes, as a physician who writes them and as an informaticist who governs the templates that produce them. The intervention is not *more CDI programs and staff*. It is template redesign, ambient and voice-recognition deployment with proper review, structured-data capture in the moments where structure matters and prose everywhere else, and clinician-led governance that treats the note as a clinical instrument first and a billing instrument as a downstream consequence.

Where that has been done, the financial lab values respond.

The Critical Lab Value of Quality-Linked Payment

A growing share of Medicare and commercial payment is contingent on documented quality. Hospital value-based purchasing, the Hospital-Acquired Condition Reduction Program, the Hospital Readmissions Reduction Program, Merit-based Incentive Payment System (MIPS) on the ambulatory side, and Promoting Interoperability all impose adjustments that can run positive or negative based on measured performance (CMS, 2024). The aggregate swing for an average health system, across all programs, can reach between 2% and 9% of total Medicare reimbursement.

For a system at a 1% operating margin, that is the difference between solvency and insolvency.

What is rarely appreciated outside the informatics community is how directly those adjustments flow from EHR configuration. The numerator and denominator of nearly every quality measure driving a payment adjustment are built from EHR data. If the order set lacks the right structured element, the measure cannot find the care that was delivered. If the documentation template does not prompt the right clinical element, the care may not be recorded at all. If the audit logic reads from a workflow that is not used anymore, the dashboard reports compliance that does not exist, leadership accepts the false reassurance, and the penalty arrives a quarter later as a surprise.

Recovering quality-linked revenue is therefore not primarily a finance project. It is an informatics project. The leader who governs decision support, order-set design, structured-data capture, audit logic, and clinical workflow is the same leader governing whether the value-based adjustment lands positive or negative.

Organizations that understand this connection invest accordingly. Organizations that do not are surprised twice a year.

The Critical Lab Value of Regulatory Penalties

Regulatory exposure is a comorbidity of delivering healthcare in America. Like any comorbidity, it can be managed, optimized, or left alone until it acutely causes pain.

Failure to meet Promoting Interoperability requirements can reduce Medicare payment updates by up to 75% of the annual market basket increase, a penalty purely informatics in origin and one that often goes uncaught in organizations that have never staffed the role (CMS, 2024). Information-blocking violations under the 21st Century Cures Act carry civil monetary penalties of up to $1 million per violation for the health IT developers, exchanges, and networks. Providers found to have committed information blocking face a separate set of disincentives finalized in 2024: exclusion from the Shared Savings Program for at least a year, a zero score in the Quality Payment Program's Promoting Interoperability category for physicians, and public posting of the violation (ONC, 2024). The Trusted Exchange Framework and Common Agreement (TEFCA) have turned interoperability from an aspiration into a measurable, auditable, penalizable capability.

The regulatory architecture is moving from theoretical to concrete enforcement at exactly the moment most provider organizations are least prepared for it.

These obligations are usually not pure IT problems. They are more often operational leadership and clinical informatics governance problems. They require a leader who understands what is being asked, what evidence demonstrates compliance, which workflow and configuration choices satisfy the requirement without burying the physician, and how to design the audit trail that will be there when Centers for Medicare & Medicaid Services (CMS) or the Office of the National Coordinator for Health Information Technology (ONC) asks for it.

Without that leader, regulatory posture is reactive, fragile, and perpetually one audit away from a public failure.

THE DIAGNOSTICIAN'S LAB REVIEW

There is a temptation, in finance, to answer every abnormal financial lab value with a financial intervention. Margin compression invites cost reduction. Denial volume invites appeal. Quality penalties invite a consultant project. Regulatory exposure invites compliance staffing. Each one has a place.

None of them is the *upstream* intervention.

The upstream intervention is to govern the instruments from which the financial lab values are eventually computed: the record, the templates, the clinical content, the structured data, the audit logic, and the workflows that produce all of it. That is *clinical informatics* work. It must be led by a physician leader who understands both the clinical content and the machinery, supported by a finance partner who understands the downstream consequences, and protected by a chief executive officer (CEO) willing to invest in the capability rather than the symptom.

Where this has been done, the lab values respond with a consistency that is difficult to ignore. Documentation completeness rises. CDI yield improves without coercion. Quality-linked adjustments turn from negative to positive. Regulatory posture becomes proactive. Denials fall. Margin recovers, and not because the organization is billing more aggressively. It recovers because the specimen finally reflects the care being delivered.

BEDSIDE QUESTIONS: THE EXAMINATION OF YOUR ORGANIZATION

1. Can you trace any one of your top three quality-linked payment metrics from the bedside data element back to the dashboard, and do the workflows that produce that element actually exist?
2. Is your CDI program integrated with the EHR at the workflow level, or running as a parallel chart-review function?
3. Is your clinical content designed to capture the structured elements your payers' automated systems require?
4. Who owns Promoting Interoperability, TEFCA participation, and Cures Act compliance? Is it a clinical informatics leader, or has it been absorbed into IT compliance?
5. What is your organization's total annual spend on health information technology, counting licensing, implementation, consulting, cybersecurity, analytics, and

AI subscriptions? Can you produce that number as a single figure?

6. When was the last time your Diagnostician (CMIO or equivalent) and your CFO sat in the same room with a single, joint agenda?
7. Is documentation in your organization governed as a *clinical* instrument or as a *billing* instrument? The answer is rarely both, and the choice has financial consequences.
8. If a payer's denial pattern shifts tomorrow, who is responsible for diagnosing the cause within 30 days?

If the answer to these questions reveals bad news, or no news, you are reading your own lab values without a Diagnostician.

THE LAB VALUES HAVE BEEN DRAWN, AND THE PATHOLOGY GROWS clearer with every number on the report. But every finding in this workup will be recorded in an EHR, and that instrument has not yet been examined. An imaging study is only as useful as its data and the skill of the person reading it. If the reader cannot distinguish artifact from anatomy, it may be worse than having no study at all.

CHAPTER 4
THE EHR
AN IMAGING MODALITY FOR INFORMATION

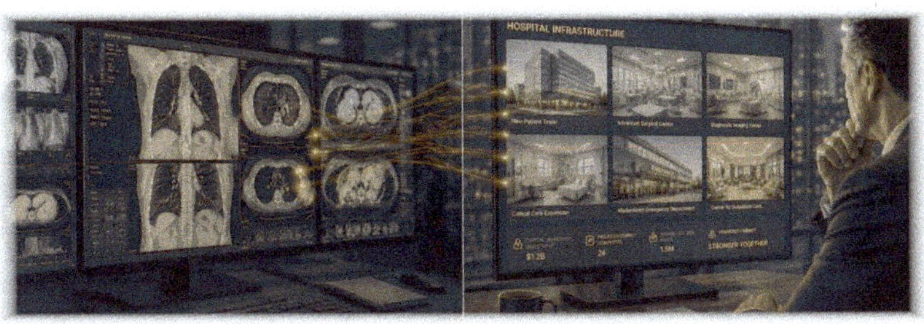

A MAGNETIC RESONANCE IMAGING (MRI) DOES NOT HEAL A PATIENT. It produces images. A skilled radiologist, partnered with a skilled physician, turns those images into diagnoses, and the diagnoses inform a treatment plan. The EHR works the same way, but the organizations that have treated it as the treatment rather than the imaging study have learned, slowly and expensively, that an MRI without a radiologist is just an enormous magnet.

A 37-YEAR-OLD MAN VISITS HIS FAMILY PHYSICIAN WITH SIX WEEKS OF worsening low back pain, starting while lifting a heavy box during a recent move. The pain has been getting worse by the day. Recently, the pain has begun traveling down his right leg. On exam, the physician finds some muscle weakness in that leg, some reduced feeling, and a reflex that is different from the left leg. Concerned this could be a surgical problem, the physician orders an MRI of the lower spine and refers him to a spine surgeon in town. The patient does not want surgery, but he wants the pain to go away.

Later that day, a radiologist reads the MRI study, measures the disc herniation, checks it against the nerve it may be compressing, and dictates a report. The findings reassure the spine surgeon, who matches them against their exam findings and writes the treatment plan: physical therapy first, an injection if that fails, surgery only if the issues persist.

The MRI itself is not the treatment. It does produce the important images that will guide the diagnosis and course of care that follows.

INTRODUCING THE EHR

In 2009, when the HITECH Act began seeding the national investment that would push virtually every U.S. hospital and most ambulatory practices onto certified EHR, a generation of healthcare leaders made a small but consequential error. They began, almost universally, to refer to the EHR as *a solution*. The right EHR would solve the quality problem. The right EHR would solve the safety problem. The right EHR would solve the burnout problem. The right EHR would solve, eventually, every problem that healthcare leaders had been told the right EHR could solve (Washington et al., 2017).

The prediction was wrong because the category was wrong.

The EHR is an imaging modality. It produces a digital image of the clinical encounter, including the orders placed, medications administered, vitals recorded, notes written, results received, and decisions made. It renders that image visible across time, across providers, across settings, across enterprises. Like every imaging modality, it is enormously powerful when read by a Diagnostician who knows how to read it, and nearly useless when treated as the treatment itself.

My warnings can be misread as criticism of technology or IT. They are not. I have been a technophile my whole life. But I have also been in this industry long enough to know that IT and clinicians are operating in two completely different paradigms. The reality is that our technology departments are fiercely protective of their systems and fully invested in the belief that the EHR is the ultimate solution. But we must be honest about boundaries. An IT tool and IT background do not provide the clinical intuition required to design a safe patient workflow. Expecting clinicians to completely yield their data, workflows, and decisions to a technical system is a structural design flaw that compromises care.

I am also a career chief medical information officer. I have spent my career inside these systems. I have watched them mature into something genuinely powerful, and I recommend their thoughtful adoption to every organization I advise. The electronic health record is a good tool and a durable one that is not going away.

But it is a tool, and this chapter is about the difference that word makes. The organizations that struggle with their EHR are almost never the ones that bought the wrong product. They are the ones that expected the product to do the work that only leadership, governance, and sustained effort can do.

A good instrument does not fix a struggling organization. The work does.

What follows is how to use the imaging modality the way a physician uses any other: with a clear indication, with attention to image quality, with skilled interpretation, with an awareness of artifacts, and with a plan for what to do with what is found.

WHAT THE EHR IS AND IS NOT

The cardiac MRI machine is a remarkable instrument. It identifies myocardial fibrosis, characterizes the heart's valve and chamber sizes, and visualizes blood flow in three dimensions. It costs millions to install, requires a physical environment engineered around it, demands ongoing maintenance, and consumes a disproportionate share of the imaging department's capital budget.

Hospitals do not buy the cardiac MRI because they want a *magnet*. They buy it because the magnet, in the hands of a trained radiologist working with the referring physician, produces information that changes patient outcomes.

That magnet without a radiologist is just a very expensive paperweight. A radiologist without a referring physician is a fluent reader with no question to answer. And, a referring physician who does not understand what the magnet can and cannot do will order the wrong study, ask the wrong question, misread the finding, and commit the patient to the wrong plan.

The EHR is the same. A $200 million EHR without a Diagnostician to read it is paying for the magnet and skipping the radiologist. A clinical informatics leader with no clear clinical question to answer is fluent in the imaging with no indication for it. And, a leadership team that does not understand what the EHR can and cannot do, what questions it answers, how the image is acquired, and where the artifacts come from will commission the wrong study, accept the wrong reading, and act on the wrong picture.

The EHR can do extraordinary things. It can render the longitudinal arc of a chronic disease visible immediately. It can surface the right alert at the right moment and prevent a fatal interaction. It can drive evidence-based pathways into the rhythm of clinical care. It can make documentation a side effect of practice rather than a separate event. It can route a critical result to the right inbox at the right urgency and confirm the loop is closed. It can do all of that when it is configured by people who understand the clinical questions, the technical instrument, and the way the work occurs.

IT CANNOT DO any of it if the clinical questions are missing.

HITECH AT SEVENTEEN

Most will agree that seventeen years is long enough to evaluate any treatment. By this point in a patient's course, a physician no longer cares whether the drug was well intentioned. They ask whether it worked, and if it did not, *why*.

It is time to hold the electronic health record to that standard, because too many good people spent seventeen years believing it was the treatment rather than the imaging study.

Begin with what it costs. A federal incentive program that paid hospitals up to roughly $6 million, and individual physicians up to $44,000, financed systems that now routinely cost large health systems more than $200 million to implement, and at the largest academic and multi-state systems, well over $1 billion. The build is not the end of the bill. Licensing, maintenance, optimization staff, and the specialized workforce needed to run the platform well typically run eight figures a year, every year, for as long as the organization exists.

The incentive was the down payment. The mortgage is permanent.

The Missing Measure

Every organization in America can tell you what its EHR costs. Almost none can tell you what it returns.

That asymmetry is not an accounting oversight. It is the natural result of booking health information technology as an expense category, which is how every finance department in the country currently treats it. Expenses get minimized. Instruments get read. The two disciplines look nothing alike, and healthcare has spent seventeen years confusing them.

As discussed in Chapter 3, health information technology is roughly 1% of national health expenditure and about 2% to 3% of a hospital's operating budget. Administration, the largest single category of avoidable cost in American healthcare, runs at 34%. And, the

1% category is the instrument through which the 34% category is either amplified or reduced. Every prior authorization, claim, denial, appeal, coded diagnosis, and quality measure passes through it.

So, the useful question is not what the *EHR costs*. It is what the *EHR returns* against the overhead it touches.

That measure does not currently exist on any hospital's dashboard, and I would propose it as the target performance measure of the next decade. Not technology spend as a percentage of operating expense. Not go-lives completed on schedule. Not user satisfaction scores collected six weeks after cutover.

RETURN ON THE INFORMATICS FUNCTION, MEASURED AGAINST THE administrative and clinical burden it was purchased to reduce, tracked over years rather than quarters, and owned by a named executive.

An organization that measures cost will always underinvest in the instrument. An organization that measures return will discover, usually within two budget cycles, that it has been starving the highest-leverage line item it owns.

Every dollar inside the permanent EHR run-rate is also a dollar not spent on a nurse-to-patient ratio, a clinical hire, or the informatics workforce this book argues every organization needs. A board approving next year's budget should ask, in the same meeting, what this money bought, and what else that money could have bought. Not as a rhetorical exercise.

What the Market Is Doing

It is important to understand the market you are selecting within. For most of the seventeen years since the incentive began, an organization large enough to run a full hospital encounter, admission through discharge, order through result, chart through claim, has had in practical terms two platforms able to do it at enterprise scale. This is the operating reality every Diagnostician who has run a vendor selection in the last decade has lived through.

The market has continued to consolidate. As of this writing, one of the two enterprise platforms is steadily gaining share while the other is losing it, and conversions from one to the other that were unthinkable a decade ago have become routine (KLAS Research, 2026b). A market that began as a genuine, if narrow, competition between two capable platforms is settling toward something closer to one. This matters, because it shapes your negotiating leverage, your timeline, and your assumptions about where the field will sit when your next contract comes due.

It is also worth noticing how slowly it has moved, because the reason is instructive. The same switching costs and network effects that built this concentration are why enterprise platforms do not simply collapse. An organization on any enterprise platform faces the same multi-year, multi-hundred-million-dollar migration whether it is leaving a struggling system or joining a thriving one. The forces that narrowed the market are the same ones that make any exit from it slow and expensive.

Whether any of that predicts how an organization performs is a separate question, and it has been studied carefully. *The Natural Experiment*, later in this chapter, reports what seventeen years of measurement found.

CONFUSING THE TOOL FOR THE TREATMENT

Vendors are pitching U.S. health systems with increasing intensity in 2026: that adopting one of the leading electronic-health-record platforms is, substantially on its own, an organizational fix.

The case deserves a fair hearing, and the smaller or resource-constrained system should take it especially seriously. A community hospital or a modest regional system genuinely cannot build on its own what a platform's scale delivers: build patterns refined across a decade of enterprise installs, the depth of a trained implementation workforce, and the interoperability of a large, connected network. For an organization that could never assemble those things independently, buying access to them is a rational and possibly necessary decision.

The pitch, like every pitch, is made or reinforced by an interested party. The vendor is a company with its own agenda and business model, including durable revenue, expanding share, and a defensible position, and there is nothing improper in that. That is what a company is for.

A pharmaceutical representative is professionally obligated to present a drug in its best light. The obligation to weigh that drug independently, against the patient in front of you, sits with the prescriber. The same division of labor applies here. The vendor's job is to sell the platform well. The executive-physician's job is to read the pitch the way a Diagnostician reads any therapeutic claim, with the patient's condition in view rather than the sales decks.

One feature of the client relationship is worth understanding plainly, because it is rarely discussed in the sales process and it shapes everything downstream. EHR platforms are currently considered software, and software is lightly governed. These systems sit largely outside Food and Drug Administration (FDA) oversight, carry only limited Office of the National Coordinator for Health Information Technology (ONC) obligations, and routinely include broad hold-harmless provisions in their contracts.

The practical effect is an asymmetry the executive should walk in already understanding. The organization carries the patient-safety mission. The clinician carries a license at the bedside. The facility carries the regulatory and legal exposure. The vendor, by contract, carries little of this downstream liability. When a mis-tuned alert, a flawed order set, or a missed regulatory capture contributes to harm, the responsibility to own it, defend it, and remediate it sits with the organization.

None of that is hidden, and it reflects a national regulatory framework that was poorly thought out. It is simply the shape of the deal, and it is one more reason the point this book keeps returning to holds. The organization, not the vendor, is accountable for what the tool does in its hands. One of the executive-physician's earliest jobs is to make sure the organization's interests, and its patients', are represented in the contract and in the build, because the sales process will not represent them for you.

What a platform vendor sells is access to a foundation. The implication is that the rest of the work follows from a strong foundation the way a house follows from a well-poured slab, and that metaphor is more right than wrong. A foundation matters.

What a foundation does not do, in construction or in this argument, is build the house. And, a foundation poured to someone else's specifications is unlikely to be the right fit for what your organization is building.

THE HONEST ACCOUNTING

Here the executive needs a real accounting, and, unfortunately, it is not a ledger with the positives on one page and the negatives on another. Nearly every genuine strength an EHR platform brings currently carries its own limitations. The Diagnostician's task is to hold the strengths and the limits in the same hand. Let us discuss five of them.

Network effects, and their real reach.

When a substantial share of the patient's prior care, the referring physician's record, and the community's other providers all sit on the same instance of the same underlying platform, longitudinal data fluidity and continuity are materially better than they are across differing platforms. The arriving record is more often complete. The handoff at care transitions is more often clean.

The word doing the work in that sentence is *substantial*, and in most American markets it is not the reality. The reality is that most independent ambulatory practices, behavioral-health networks, specialty groups that were never acquired, and skilled-nursing and post-acute partners are running on smaller EHR systems built for different economics. They remain common enough in most communities that a genuinely single-platform patient record is closer to an aspiration than a reality.

A poorly understood part of our reality is that where organizations run the exact same vendor platform, *the same platform* does not mean that they seamlessly share information. Two instances of the identical product, configured by two different organizations over two different implementation histories, routinely diverge in problem-list vocabularies, order-set logic, and local data dictionaries. Cross-organization exchange, even between same-vendor instances, typically arrives as a shareable document rather than as discrete, mapped data the receiving system's decision support can act on without translation.

The network effect is real and worth having. In most 2026 deployments, it is still not the frictionless interoperability the pitch implies.

Workforce depth, and what it does and does not produce.

The pool of clinicians, IT analysts, build engineers, and trained informaticists fluent in our leading platforms is larger by an order of magnitude than the pool fluent in others. Recruiting is easier. Onboarding is faster. The cost of replacing a departing staff member is lower because the next one already speaks the language. The advantage is real and worth banking.

The caveat rests completely on the sophistication of the purchaser. Fluency in a platform's build tools produces IT analysts, users, and support staff who can install and configure the system. It does not produce informaticists who can optimize and govern it. The judgment this book has been arguing for is not taught by a vendor certification program, because it is not a *platform skill*. It is a *clinical skill* that happens to be exercised inside of a platform.

A deep bench of certified builders makes it easier to execute a governed optimization program. It does not create one. Organizations that mistake these two are the same organizations that eventually discover their roadmap has quietly become the vendor's roadmap, and that the original diagnostic question, "*What clinical priority does this serve?*" has been replaced by a different one: "*What does the next release let us turn on*?"

Build maturity and the median it was built from.

Two decades of experience, accumulating from thousands of deployments, have produced a library of build patterns that have been refined repeatedly by the vendor and its user community. An organization deploying on a platform is rarely starting from scratch, and the base configuration, while never adequate as a final state, is usually a defensible starting point.

This is the strength most often over-read.

The library reflects the aggregate experience of thousands of other organizations' patient populations, payer mixes, and specialty distributions. *Not yours*. And, this book has already shown that the median performance of modern American healthcare is not a strong target to aim at. A foundation engineered to the median of a system that has issues is not, by virtue of its scale and polish, engineered to excellence.

It is engineered to consensus.

Those are not the same achievement, and the difference is the argument of this book.

The organizations that benefit most from build-pattern maturity are the ones that audit the pattern against their own population before adopting it, not the ones that adopt it because auditing takes longer. What happens when the pattern does not get audited is measurable, and it is the subject of *The Natural Experiment* below.

Investment scale, and who actually benefits.

The leading vendors invest in research, development, and emerging capability, including AI and the evaluation machinery around it. This is a scale that individual health systems cannot match. This is where every client's case is strongest, especially the smaller organizations. A rural hospital or an independent regional system gains access to capability it could not plausibly build for itself at any price. That is not only a marketing claim, but also the single best argument for adoption a resource-constrained system has.

It is also not the whole story. The gap between capability shipped and capability used is the part boards underweight. New functionality arriving in a release note is not necessarily new functionality implemented, governed, validated against the local population, and adopted into the clinical workflow.

The organizations best positioned to fund that last step are disproportionately the large, well-resourced systems that needed the vendor's scale the least. The smaller system gains the access. It does not automatically gain the informatics capacity to safely turn it on, and the capacity is the part that determines whether access ever becomes value.

Integration and the savings it produces.

A fully consolidated platform reduces the burden of integration. Separate interfaces, identity reconciliation, and terminology mapping consume a substantial share of an IT budget in more complex environments. The savings are real and recurring.

For most organizations, they are also still mostly theoretical. Laboratory reference systems, PACS and imaging archives, revenue-cycle add-ons, behavioral-health and oncology specialty systems, and the ambulatory practices absorbed through a decade of acquisitions tend to persist alongside the core platform. Each of these requires its own interface to maintain. *Fully consolidated* is the operative qualifier, and few real health systems have reached it.

The savings that are real come from the same structural fact that produces this chapter's most serious concern. A platform earns its efficiency by becoming the single vendor an organization runs on. By construction, it also produces vendor dependencies.

What that dependency costs is the pitch's other half.

ONE CONDITION, THREE SYMPTOMS

The dependency created by a single vendor relationship creates three symptoms that feel like three separate problems. They are, however, one condition: the absence of a credible alternative, showing up in three places that are easy to budget for separately and dangerous to think about separately.

The first symptom: the financial and structural exit cost

Once an organization has built years of order sets, interfaces, workforce training, and institutional muscle memory around a single platform, changing platforms is a multi-year, multi-hundred-million-dollar undertaking rather than a procurement decision. A customer with a credible alternative has negotiating leverage. A customer who has structurally committed to one vendor does not. The pricing, terms, and roadmap all drift in the direction that dependency suggests.

This is also the trap that has kept the vendor market narrow for seventeen years, and it was named as a trap before most of these contracts were signed (Mandl and Kohane, 2012). The cost of switching platforms protects them from being abandoned, *because abandoning a struggling platform is as expensive as staying on one.*

The second symptom: security exposure

When a small number of platforms hold most of the country's hospital records, a successful attack against any one of them stops being a local incident. It becomes a national-scale exposure.

This is not hypothetical. In 2024, a single ransomware attack against one national claims-processing vendor froze prescription routing, insurance verification, and billing for thousands of pharmacies and health systems simultaneously, for weeks. The event had nothing to do with any individual hospital's own security posture and everything to do with how much of the country's health-financial plumbing ran through one company (Kanter, 2024).

A parallel event at either dominant platform, now connecting most of the nation's clinical records, would not be a story about one company's breach. It would be a story about how much of American medicine stopped documenting, ordering, and billing on the same afternoon.

A recent peer-reviewed analysis describes exactly this as a genuine tail risk of EHR consolidation, low in probability and catastrophic in consequence, compounded by every hospital's growing dependence on the same handful of infrastructure providers (Holmgren, Apathy, and Kanter, 2025).

The third symptom: quality-and-safety exposure

The biology metaphor is the clearest way to see it. A field of genetically identical crops is highly productive, and uniquely vulnerable, because a single pathogen that defeats one plant can defeat all of them. A clinical-care ecosystem where most organizations run through the same platform is both similarly productive and similarly vulnerable; not only to the kind of event described above, but also to the proliferation of a flawed clinical decision support (CDS) rule, a misconfigured default, or a regulatory misalignment in the platform's logic. Any of those now carries population-scale consequences, because the platform is the substrate for population-scale clinical activity.

The leading platform's operational reliability is, on average, excellent. Its exposure risk is structural.

All three of these symptoms trace back to the same root, which is why treating them as unrelated items understates the situation.

The EHR consolidation that makes the content robust also makes exiting expensive, and makes a single security or misconfiguration event national in scope. These are outcome risks at population-scale.

The authors studying the security exposure make a point worth acknowledging. Reversing this consolidation is not realistic, given the cost of switching already described. The concentration is not going away. What can still change is whether the organizations and regulators depending on these platforms treat the risk as known, and manage it. Organizations and regulators should seriously consider redundancy, incident planning, contractual protection, and governed exit-cost discipline. These are significant risks.

WHAT THE PITCH COSTS

Underneath every symptom above sits a fact this book has already established. The platform changes none of the underlying tenets, just the tools. In the hands of an organization without a properly resourced clinical informatics program, the dominant platforms produce the same alert override rates, the same note bloat, the same dashboard artifacts, the same iatrogenic safety patterns, and the same workforce burden as any other platform. The foundation configuration is just more sophisticated. The absence of governance produces the same downstream pathology.

Buying a leading platform without building the Diagnostician Function is empiric treatment without diagnosis.

The largest effect of buying the pitch is what the pitch does to the organization's attention and budget, before a single module goes live.

A board told that the platform is the transformation treats go-live as the destination. The Diagnostician Function does not get funded, because the platform was supposed to do that work. CDS governance is not chartered, because vendor defaults were supposed to be adequate. Data-quality discipline is not built, because the platform's reporting was supposed to be sufficient.

This is the foundation problem in its most seductive form. *The organization forgets the house still must be built.* The foundation looks strong enough on its own, and the work of the Diagnostician appears unneeded.

It is not the platform's fault that this happens, and this book is not arguing that it is. A platform cannot mandate the premises a client brings to the purchase. An EHR can only be as good as the governance and processes built around it, and the vendor has no mechanism or motivation to insist that either exist before the money flows.

But the fact that it is not the platform's fault does not make it any less expensive. Every dollar and every year of attention an organization spends believing the platform has solved the diagnostic deficiency problem is a dollar and a year not spent building the actual foundation this book has described: the executive-physician, the informatics workforce, and the governance discipline that turns a well-built magnet into a functioning instrument. The platform, marketed to bypass that investment, ends up choking it off entirely.

Three years later, the dashboards are wrong, the alert override rates are high, and the executive team is asking why the most expensive transformation in the organization's history has not transformed anything.

The diagnosis is the same one this book opened with, now bearing a *much more expensive price tag*.

THE FOUNDATION ARGUMENT

A physician faced with a marketing pitch for a new class of medication asks a familiar set of questions. What is the indication? What is the contraindication? What is the expected effect, in what population, on what timeline? What are the adverse effects and the interactions? What does the discontinuation criterion look like? Is the evidence base independent of the manufacturer?

Those are the questions a Diagnostician should ask of the platform.

Indication. Purchasing or transitioning to a new EHR is most clearly indicated when the organization has no platform or is exiting an obsolete or unsupported one. It can also be beneficial when networking with significant partners who are already on the same platform. And, it should be driven by an identified operational issue, if local clinical informatics capability is robust enough to govern the configuration, rather than passively accept vendor defaults.

Contraindication. The platform is not indicated when the migration is being sold as a substitute for the diagnostic and governance work this book describes, when the executive sponsors believe go-live is the transformation rather than the start of it, and when post-implementation informatics resourcing is being deferred or scaled down on the assumption that the platform will absorb the work.

Expected effect. A clean migration, well governed, well executed, in an organization with existing clinical informatics capacity, can produce measurable gains in workforce satisfaction, cross-organizational data fluidity, quality metric capture, and operational consistency within 12 to 24 months. A poorly executed migration produces a year of disruption followed by the same pre-migration symptoms in a slightly more sophisticated dress.

Adverse effects. Vendor capture. Roadmap drift away from local priority, and the complacency created by false expectations.

Discontinuation criterion. Platform deployments are, in practice, difficult to discontinue. The exit cost is the negotiating leverage the vendor knows it has. The discontinuation criterion is therefore theoretical, which means the indication and the governance must be exactly right at the time of adoption.

Evidence base. The vendor's marketing material is not the evidence base. The peer-reviewed literature on platform-specific outcomes, the independent analyses of comparative performance, and the direct experience of organizations that have completed the migration with and without proper governance are the evidence base. The Diagnostician reads those, not the slide deck.

THE NATURAL EXPERIMENT

Medicine rarely gets a clean natural experiment at a national scale. This field has one, and its importance has gone largely unnoticed.

Take thousands of hospitals. Give them the same class of intervention, drawn from a handful of vendors. Let seventeen years pass. Then measure.

If the software is the cure, the groups treated should be healed. If something else is the cure, they will not be. They have not been healed. *They have not come close.*

The labs that came back negative.

Independent, peer-reviewed analyses of thousands of hospitals have asked whether the vendor choice predicts performance. The vendor only explains a small fraction of the differences in quality and safety between organizations, far less than local implementation and support. It bears no statistically significant relationship to financial performance (Beauvais et al., 2021; Holmgren et al., 2021).

That is usually reported as a negative result and set aside. It should not be. It works exactly the way a negative workup works at the bedside.

The variable that was being measured, quality and safety, were not improved. Scientific and regulatory reviews show that the variation in numerous measurements between locations using the same EHR is both real, and large. Therefore, the source of that variance lives somewhere that has not been measured.

The easier path, and what it actually costs

There is an easier path, and it is genuinely tempting.

Rather than doing the hard and slow local work of evaluating platform content against your own population, an organization can adopt the vendor's median configuration as delivered. This is much faster and cheaper up front. There is a reassuring sense that thousands of other institutions arrived at the same answer, and it removes a hundred arguments the organization would rather not have.

I absolutely understand the appeal. I have sat in the meetings where it is proposed, and the people proposing it are not lazy or unserious. They are tired, and they are looking at a backlog of work they cannot staff.

But consider what your local configuration is. It is not arbitrary, and it is not simply the residue of neglect. Every local variation grew out of some real pressure. *Some of those pressures were good*: a service line that built something better than the default, a safety event that produced a hard-won safeguard, or a population whose needs genuinely differ from the national median. *Some were bad*: a loud voice that got its way, a workaround that outlived its cause, or a build that never got cleaned up.

Good and bad, they are all real and each represents previous efforts meant to help, until they are examined more closely.

Replacing that configuration wholesale, for a "foundation," is not just a technical migration; it is a decision to change almost everything the organization does. The clinical workflows change, and so does every administrative process attached to them: scheduling, registration, coding, charge capture, prior authorization, quality abstraction, and all the reporting built on top. The pain is not confined to the clinical staff, and it does not end at go-live.

A systematic review of 40 studies on moving from one electronic record to another makes the point directly. Transitioning between digital records presents different challenges than the original transition from paper, and the literature on paper tells you very little about the digital (Miake-Lye et al., 2023). Any organization still reasoning from its paper conversion is preparing for the wrong surgery. And, we are very early into the latest large-scale digital migration pattern. It is lightly traveled, but accelerating.

None of that argues against migration. Sometimes it is exactly right, and the indications above say when. It argues against migration undertaken as a substitute for the diagnostic work, because the organization pays the full price of the disruption and never receives the thing it thought it was buying.

What the national tests show

If the platforms were the cure or leveling force, organizations starting from the same configuration would end up in roughly the same place.

They do not, and we can measure it precisely, because for more than 15 years hospitals have been running a standardized safety test on their live production systems. The Leapfrog Group administers it (Leapfrog Group, 2025). The method was published two decades ago and has been applied consistently since (Kilbridge et al., 2006). It uses simulated medication orders drawn from cases that injured or killed patients and asks one question: "Does your system catch them?"

When Metzger and colleagues first reported national results in 2010, across 62 hospitals, the involved systems caught only 53% of the orders that would have been fatal. Detection of orders that would have caused serious harm ranged widely, from 10% up to 82% (Metzger et al., 2010).

A decade later, Classen and colleagues published the full national series: 8,657 hospital-years of testing between 2009 and 2018. Performance had improved, and the improvement is worth acknowledging. Overall scores rose from a mean of 53.9% to 65.6%. The response scores of basic decision support rose from 69.8% to 85.6% (Classen et al., 2020).

Then look at advanced decision support, which is where the difficult clinical judgment lives, the whole purpose of decision support. In 2018, after a decade of national investment and optimization, the mean score was only 46.1%, with a large standard deviation of 21.6. One standard deviation runs roughly between 25% and 68%. *Hospitals at the best end of the curve catch only 68% of the dangerous orders. Hospitals at the other end catch only 25%.* The authors' own conclusion is that the wide variation in safety performance persists across a large sample of hospitals and EHR vendors.

Same test. Same commercially available software. In many cases the same vendor. Two very different levels of medication safety.

The feedback from human experience

The Arch Collaborative at KLAS has spent years measuring the other half of this, surveying clinicians directly about their experience using the EHR.

The 2026 KLAS Arch Collaborative "Net EHR Experience Score" reveals that targeted optimization, rather than platform replacement, drives a **6.2-point** average increase in clinician experience scores (KLAS Research, 2026a). These are *not* organizations that replaced their platform. They are organizations that kept the software they already owned and did the hard work.

The report also exposes a significant experience gap: only 12% of nurses report an elite EHR experience compared to 22% of physicians, which shows that differences in user experience are driven by configuration, training, and workflow design rather than the platform itself. These are troublingly low scores, and the gap between them makes it worse. When you purchase a platform, you affect every organizational user. Their experiences are not equal.

THE PLATFORM VERDICT

The dominant EHR platforms are, in 2026, powerful instruments available to American healthcare leaders. Taken alone, they are not a cure for the organizational illness this book has named. Diagnostic deficiency, governance fragmentation, and the absence of the executive-physician leadership and informatics capacity are not addressed by changing the imaging modality. They are addressed by hiring the radiologist who can read the image, regardless of which manufacturer made the magnet.

The platform is an image, not a treatment.

A reader who absorbs that distinction will navigate the next decade of consolidation more competently than one who does not, because they will spend their attention where it changes outcomes. The platform helps a struggling organization, only the work fixes it, and the work is the same with either platform.

Competent navigation starts with a more basic question than which vendor to choose. It starts with "*Indicated for what?*"

THE INDICATIONS FOR AN EHR

Every prescription has indications for when and why it is used. Every imaging study also has indications. The EHR, like any diagnostic modality, is most useful when the organization has been deliberate about the indications it is meant to satisfy.

In my experience, an EHR is most reliably valuable when it is configured to support five clinical indications, each corresponding to a defined clinical problem the imaging modality is uniquely able to illuminate.

The *first* is the longitudinal record itself: the durable, queryable, multi-provider account of a patient's care over time. This is the most basic indication and the easiest to fulfill, and it is fulfilled poorly when the record is fragmented across enterprises, polluted by copy-forward, or rendered unreadable by note bloat.

The *second* is decision support at the point of care: the right evidence, the right alert, the right order set, the right document, surfaced at the right moment in the workflow. This indication is satisfied when decision support is governed by a clinical leader and calibrated against override rates, near-miss reports, and clinical outcomes. It is rarely satisfied when it is configured from vendor defaults that are never reviewed.

The *third* is structured-data capture sufficient for measurement, research, and risk adjustment: the deliberate capture of discrete elements that feed downstream business analytics, quality reporting, and risk-adjustment models. Satisfying it requires that order sets, templates, flowsheets, and transcription tools be designed around the elements needed downstream. That is a deceptively difficult engineering problem, and it demands close partnership between informatics, clinical operations, quality, and finance.

The *fourth* is safe orders and safe medication management: the structured prescription, the verified administration, the closed-loop feedback when a result is out of range. This indication is satisfied when computerized provider order entry (CPOE) is integrated with decision support, when bar-code medication administration is honored at the bedside, and when the rules and the overrides are governed.

The *fifth* is interoperability and care coordination across settings: the reliable handoff of information from inpatient to outpatient, from acute to post-acute, from organization to organization, in a form the receiving clinician can use. This indication is satisfied when the organization treats the movement of information as a clinical capability rather than a compliance obligation.

Every healthcare leader should be able to say how well each of those five is being satisfied in their own organization. Most cannot. Closing that gap is work the Diagnostician is uniquely positioned to lead. (See Figure 4.1.)

ORDERING THE RIGHT STUDY

Reading the EHR Image

The EHR is an imaging modality. Someone has to read the film.

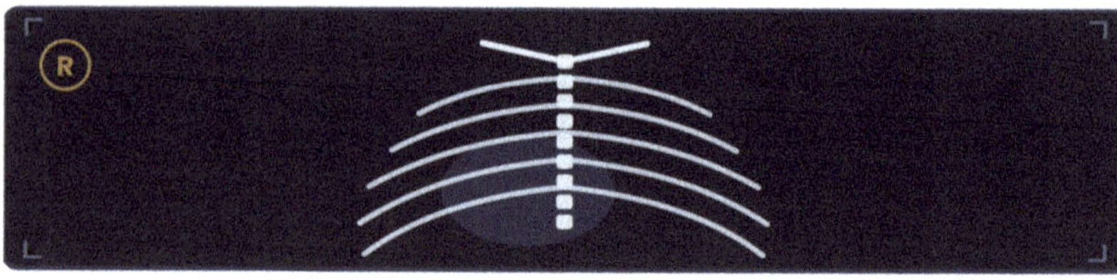

	WHAT IT REQUIRES	WHERE IT FAILS
Longitudinal Record	The durable record, spanning every provider.	Fragmented, buried in copy-forward.
Decision Support	The right alert, at the right moment.	Left on unreviewed vendor defaults.
Structured-Data Capture	Discrete data, built for analytics.	Templates built for billing, not analytics.
Safe Orders	CPOE integrated with decision support.	Overrides go unmeasured, ungoverned.
Interoper-ability	Information that follows the patient.	Treated as compliance, not capability.

A $200 million EHR without the Diagnostician to read it is paying for the magnet without paying for the radiologist.

FIGURE 4.1

Reading the electronic health record like a diagnostic image — five disciplines, one skillset.

WHERE THE IMAGE GOES WRONG

Imaging artifacts are the radiologist's recurring problem. The patient moved. The contrast did not bolus correctly. The artifact obscures the underlying anatomy, and every radiologist knows to detect and account for it. The next step is to read the image around it and decide what additional study, if any, will resolve the ambiguity.

EHR artifacts are pervasive and routinely missed. The discharge time on the audit log was not the time the patient left. The medication-administration record reflects when the nurse documented, not when the medication was given. The problem list still shows a diagnosis resolved 2 years ago. The medication list holds three duplicate entries and one drug the patient stopped taking last decade. The "vital signs reviewed" attestation was clicked because clicking was easier than scrolling. The progress notes repeat yesterday's notes nearly verbatim. The "education provided" checkbox was a default in the discharge order set and was never changed.

None of those are technology failures. They are the remnants of those who designed the workflow, who never observed it, and never used it.

They are pervasive. They corrupt the image. They mislead the analyst. They distort the dashboards. They feed the algorithms a later chapter will examine.

A leadership team without a Diagnostician is, in this analogy, a clinical service without a radiologist. The images are still being produced. They are even being read, by analysts, by report writers, by regulators, and by payers. But, there is no accountable role in the building whose job it is to recognize the artifact, account for it, and see that the next image is acquired more cleanly.

OPTIMIZATION AFTER IMPLEMENTATION

Many healthcare organizations believe they are further down the maturity road than they are. Optimization is a word spoken often and rarely defined well. Most healthcare organizations have spent the last decade in *implementation* and *stabilization*. The vocabulary is familiar. Go-live. Cutover. Big bang. Stabilization. The mental model has been one of moving from a paper and tactile process to a digital one, and the question throughout has been whether the digital one is *working*, by which most technical leaders mean *not crashing or making things worse*.

The next decade is the decade of *optimization*. Making the tools and processes optimal for the task. The vocabulary is less familiar and is usually misapplied to day-to-day EHR maintenance. This real work is harder. The governance demands are larger. The analytical sophistication required of the informaticist is higher. And, the question is no longer whether the system is working but whether it is delivering optimally on the five indications above.

Optimization is fundamentally a clinical informatics problem. It requires a leader who can sit down with the cardiology service for an afternoon and rebuild the order set against current evidence. Who can sit down with the inpatient nursing staff and redesign the flowsheet, so it captures what quality measurement requires without doubling documentation time. Who can sit down with the revenue cycle team and identify the structured captures that reduce denial volume without changing clinical practice. Who can sit down with the COO and redesign the discharge workflow, so the EHR supports the actual choreography rather than imposing a new, theoretical one.

This work cannot be outsourced. The vendor will not do it. The analyst cannot do it. It can only be done by a clinical leader fluent in both the clinical content and the technical instrument, with the time, the team, and the authority to work over months and years rather than weeks.

That is the work the modern Diagnostician does. Where it is done well, the EHR moves from being the most expensive paperweight in the organization to becoming the most powerful clinical instrument it owns.

Where it is done poorly, or not at all, the organization keeps paying the magnet's bill.

BEDSIDE QUESTIONS: THE EXAMINATION OF YOUR EHR

1. Can you state, for each of the five indications above, how well your EHR is currently satisfying it?
2. Who governs the optimization backlog? Is it clinically led, embedded in IT project management, or not happening at all?
3. When was the last time a clinician-driven change request was implemented within 90 days of submission?
4. What is your alert override rate, and who is accountable for it?
5. What proportion of your clinical content has been reviewed for evidence alignment in the last 24 months?
6. Have your audit logs and dashboard variables been audited against the actual workflow they purport to measure?
7. Does your organization book health information technology as a cost to be minimized or as an instrument to be read? Which line on the budget reflects that answer?
8. What has your total EHR run-rate returned in measurable clinical, operational, or administrative outcome over the last 3 years, and who in the organization can produce that number?
9. Do you treat your EHR as a tool, a thing you bought, or as

> an imaging modality, a capability you read? The answer determines whether it pays you back.

Unsatisfactory answers show that your organization likely has an EHR problem and no Diagnostician to bring an answer. The fate of your clinical performance, your workforce, and your margin rests on how well the instrument is read.

THE IMAGING MODALITY IS UNDERSTOOD. What has not yet been examined is the quality of the specimen it works from. Every finding in this chapter assumed the underlying data were sound. That assumption deserves the same scrutiny as the instrument, because a study read perfectly from a corrupted sample still produces the wrong diagnosis.

CHAPTER 5

HEMOLYZED SPECIMENS

WHEN THE DATA ARE CORRUPT

Every physician has been taught, from the first weeks of training, that a contaminated specimen will mislead the diagnosis. The same is true of every dashboard, every quality metric, every value-based-payment calculation, and every AI model in your organization. If the specimen is contaminated, the diagnosis will be too.

A physician trying to characterize a confusing electrolyte abnormality runs the third potassium of the morning and gets a value of 8.2 mEq/L. The patient is asymptomatic. The EKG is unchanged. The first two values, drawn from different sites, are 4.1 and 4.3 mEq/L.

The physician, trained to ask the right next question, calls the laboratory. The technician confirms what the physician has begun to suspect. The third specimen shows gross hemolysis on visual inspection. The 8.2 is not the patient's serum potassium. It is the potassium released into the tube when the red cells were sheared by a small-bore needle and a slow draw. The clinical reality is unchanged. The laboratory value in isolation is not.

What the physician does next is the diagnostic discipline this chapter is about. The physician does not re-treat the patient on the new value, does not split the difference among the three, and does not order a detailed investigation. The physician notes that the specimen was hemolyzed, identifies the value as artifact, and continues the original plan. In a well-run service the physician also notes the recurring problem with that phlebotomy team and asks whether a process needs fixing.

The artifact has been recognized. The data have been corrected for. The diagnosis has been preserved.

HEALTHCARE DATA

The above discipline must also be applied to the data on which healthcare executives make decisions every day. In most organizations it is not. The quality of healthcare data is worse than most executives assume, and the assumption itself is the vulnerability: a leader who believes the data are clean has no reason to build the discipline that catches an artifact before it becomes a decision.

Quality dashboards are read uncritically or not at all. Operational metrics are reported as though their underlying definitions were stable across departments. AI models are deployed against training data which may not match the local patient population and be catastrophically flawed. Value-based-payment calculations are accepted from CMS as though the upstream data feeding it were perfectly clean.

Virtually none of those data are perfectly clean. Almost all of them are corrupted. The question is whether the organization recognizes the artifacts.

WHERE HEALTHCARE DATA GOES WRONG

I like to call the pathway that data travel from the bedside observation to their use the ***data proposition***. Every step in the pathway is a link in a fragile chain: from the moment an observation is entered at the bedside, through its electronic capture and mapping, to guiding patient care decisions, then to aggregation, interpretation, and finally to operational outcomes.

This chapter is the examination of that chain.

A clinician observes a condition at the bedside. The observation is either captured in the EHR, or captured partially, or not captured at all. What is captured gets mapped to a discrete data element, or it sits in an unstructured text "blob" where downstream analytics cannot find it. The discrete element is extracted by an analytics query, a vendor's quality engine, or a CMS reporting interface, each of which may interpret it differently than the clinician intended. The element is aggregated, calculated, and rendered as a dashboard element, a payment adjustment, or a recommendation.

By the time an executive sees the number, it has been through six or eight transformations and is only reliable if every step was performed correctly.

In most organizations, that pathway has never been followed and audited. Each step is owned by a person who knows their own step well and the steps on either side of it poorly. The clinician does not know how the data they enter are mapped. The analyst does not know what the clinician was thinking. The dashboard designer does not know what the analyst's math is doing. The executive does not know any of it.

The number gets read as if it is a calibrated laboratory value but is closer to a many-times-transformed estimate.

The Diagnostician's responsibility is to know the path, to see that every step has a named owner, to audit the transformations against truth, and to detect an artifact before the artifact misleads the diagnosis. In my experience this is the key underdeveloped capability in modern healthcare, and one of the highest-yield places an organization can invest.

DEFINING HEALTHCARE DATA QUALITY

To clinicians, "*data quality*" often sounds like an ethereal concern belonging to a technology office. The truth is that clinical documentation is the lifeblood of the entire enterprise and one of the organization's most valuable assets. Every charted symptom, timestamped order, and coded diagnosis radiates outward to power the entire healthcare ecosystem.

Beyond the immediate bedside care, which is called the *primary use of healthcare data*, high-quality data are also used as the raw fuel for an entire array of industry evaluations, called *secondary uses of those data*. Some secondary use examples are: *clinical research* that accelerates drug discovery, identifies ideal candidates for clinical trials, and powers retrospective studies that uncover new standards of care; *population health* that tracks disease outbreaks, identifies high-risk patient cohorts, and maps social determinants of health to target preventative care; *operational efficiency* that minimizes emergency department wait times, optimizes operating room scheduling, and forecasts staffing needs to prevent nurse burnout; *precision medicine* that trains AI models to predict sepsis hours before symptoms appear and matches genetic profiles with tailored oncological therapies; *financial integrity* that guides accurate revenue cycle management, fewer insurance claim denials, and reimbursement rates that hold up under value-based care contracts; and *regulatory compliance* that satisfies mandatory federal quality reporting metrics, substantiates legal defense in malpractice cases, and maintains hospital accreditation.

There is an entire multibillion-dollar global industry formed around the aggregation, refinement, and purchase of this information. Because of its sheer complexity and profound human stakes, healthcare data have become one of the most valuable commodities in the modern economy. When a clinician ensures the data quality at the point of care, they are not just filling out a digital form; they are managing the organization's most powerful asset.

Weiskopf and Weng surveyed the field and found that assessments of electronic health record data reduce to five dimensions (Weiskopf and Weng, 2013).

- **Completeness.** Is the value present at all?
- **Correctness.** Is the value that is present true?
- **Concordance.** Do two sources that should agree about the same fact actually agree?
- **Plausibility.** Is the value believable given everything else known about the patient?
- **Currency.** Is the value still true now, or was it true once and never updated?

Kahn and colleagues later harmonized the competing vocabularies into a single framework and addressed the concern that matters most to an operating executive.

Data are not good or bad. Data are fit or unfit for a stated use (Kahn et al., 2016). The same medication list can be adequate for refill management and inadequate for a drug interaction check. Asking whether your data are good is the wrong question. Asking what you intend to do with them, and whether they are fit for that purpose, are the right ones.

Both papers carried a warning label. Weiskopf and Weng found little consistency and generalizability in how organizations assessed their own data quality, and recommended that when they reuse EHR data for decision-making, they also adopt validated, systematic methods rather than improvising.

A decade later, Lewis and colleagues repeated the review across 103 papers published between 2013 and 2023 to see what had changed (Lewis et al., 2023). The five dimensions held. So did the fragmentation.

The most commonly assessed dimension was completeness, followed by correctness, concordance, plausibility, and currency, in that order. That ranking is not a coincidence, it correlates to the cost and effort required to validate the data.

Completeness is the cheapest thing to check, because you can count populated fields without knowing any medicine. *Correctness* requires a reference standard. *Concordance* requires two systems and a person who understands both. *Plausibility* requires clinical judgment. *Currency* requires knowing what the workflow does today rather than what it did when the query was written.

So most organizations only check whether the box is filled in. Few ever check whether the box is true.

A hospital that cannot say which of the five dimensions its dashboards have been tested against has not tested them.

THE WAYS HEALTHCARE DATA FAIL

The places where data go wrong are not infinite. In my experience they cluster into six recognizable patterns. Five of them are the dimensions mentioned above. The sixth is what happens when an AI data model inherits one or more at once.

Failure to Capture (Completeness)

The clinical observation that should have been recorded as an important concept is only mentioned in the transcription or not at all. The downstream questions cannot find it. The measure reads as low, not because the care was not delivered, but because the care could not be found. This is a frequent failure found in the gap between excellent bedside care and disappointing quality scores and reimbursement.

Santos and colleagues built electronic queries to identify inpatient hypoglycemic (low blood sugar) events among patients receiving diabetic drugs, a very reasonable question, and then validated at three tertiary-care hospitals against manual chart review (Santos et al., 2020).

The queries were superb at identifying laboratory-measured hypoglycemia. Detecting a number in a database is something computers do extremely well.

The trouble was identifying which patients were receiving insulin, because insulin administered by infusion, and insulin given by other routes, were not tied to one another as the same "insulin."

The measurement was not wrong by computation. It was blind to the sickest diabetic patients in the entire building, the ones on an infusion. The data were corrupted from their design.

A hospital reporting this measure would report a hypoglycemia rate that excluded its most vulnerable diabetic patients, and every number downstream would look clean.

Corruption at Collection (Correctness)

The clearest published example is not a software problem at all. It is a sensor.

Sjoding and colleagues compared pulse oximetry readings to arterial blood gas measurements and found that black patients had nearly 3 times the frequency of unidentified hypoxemia (an arterial oxygen saturation below 88%) while the pulse oximeter read a reassuring 92% to 96% (Sjoding et al., 2020). The device was not broken. It was calibrated on a different population.

Now follow that number through the data proposition.

The result enters the chart as a "fact." It feeds the early warning score. It informs the removal of oxygen, the ventilator use, the sepsis screen, and the intensive care unit (ICU) transfer. It is also calculated into quality measures. It is warehoused, and eventually trains an AI model.

Those downstream measures are correct in their math and wrong in their conclusion. The field was recorded. It was wrong from the start. Everything downstream is reading the same corrupted specimen.

This is the failure that organizations check for least because it is hard to do, but should fear the most.

Different Definitions (Concordance)

The element is captured, but its definition has drifted between departments, systems, or time. The discharge time recorded in nursing documents is not the same discharge time the bed-control system records, and not the same discharge time found in the EHR audit log. None of them agrees with the others, and the dashboards built on each will tell different stories.

An executive who reads them and notices the discrepancy is told that they are measuring different things. An executive who does not notice is making decisions using data that may not describe the same underlying reality.

Changing Workflow (Currency)

The question the organization is asking today is based upon a workflow that stopped existing six months ago. The order set was replaced, the documentation was redesigned, the clinical decision support (CDS) rule was deactivated, and the question knows none of this.

Changes go unnoticed, because the number still looks normal. This failure is far more common than is appreciated, and it is the one most likely to be running in your organization right now.

Singh, Murphy, and colleagues found abnormal test results reported in electronic notification systems that no clinician was reading anymore, with the design of the notification identified as a contributing cause (Singh et al., 2009; Murphy et al., 2017). The data were captured, correct, concordant, and plausible. They were simply no longer current.

Aggregation Errors (Plausibility)

The data are correct for the patients, and the math is wrong for the group, somewhere upstream: how missing values are handled, how subgroups are weighted, how the calculations treat outliers, and dozens of other small choices invisible to readers of the finished dashboard.

Aggregation artifacts are particularly dangerous because the patient-level audit confirms the data are right, while the population-level metric misleads. This failure often survives the rare audit some organizations perform.

Algorithm and AI Training-Data Assumptions

The algorithm or AI model was trained on a population that does not match the local population, or a workflow that does not match the local workflow, or an outcome that does not match the local definition. Worse, it is trained on EHR data with every data issue we have discussed up to this point. It performs as advertised on some training group and degrades upon local use.

When Obermeyer and colleagues examined a widely used population health algorithm, an assumption corrupted the result. It used a proxy (the common practice of using one measure to approximate another). Prior healthcare spending was used as a proxy to assess prior healthcare need (e.g., healthy people spend less on healthcare, so less spending means healthy). Because less money had historically been spent on black patients, as a cohort, the algorithm concluded that they were healthier (Obermeyer et al., 2019).

The arithmetic was correct. The proxy was wrong. And, we have no way of knowing what assumptions are built into the models that have never been audited. The same literature runs the other way when the design is right. Pierson and colleagues built a model that found pain in underserved patients that radiologists had been missing (Pierson et al., 2021). The variable is not the algorithm. It is whether its training target was ever governed and tested.

One corruption enters at data entry, another at the sensor, another at the math, another at the proxy, and all are invisible to users downstream because every check the organization performs comes back clean. (See Figure 5.1.)

HEMOLYZED SPECIMENS · DATA CORRUPTION

Healthcare Data Failures

Six recognizable ways clinical data goes wrong – and every check comes back clean.

#	Failure	Dimension	Description
1	**Failed Capture**	COMPLETENESS	Never entered as discrete data, so it can't be measured at all.
2	**Corrupted Source**	CORRECTNESS	Present, structured, and wrong — every number downstream inherits it.
3	**Changing Definitions**	CONCORDANCE	Same term, different meaning by system — two dashboards, two stories.
4	**Changing Workflow**	CURRENCY	The process moved on; the pipeline still tracks the one it replaced.
5	**Misleading Averages**	PLAUSIBILITY	Every patient's data is right; the population math still misleads.
6	**Mismatched Training**	INHERITS ALL FIVE	A model trained on someone else's population, applied blindly to yours.

A Diagnostician audits every dimension. Most organizations only ever check one.

FIGURE 5.1

Clinical data goes wrong in six recognizable patterns, plus the model that inherits them all.

THE DOCUMENTATION SPECIMEN

One source deserves separate treatment, because it feeds nearly everything else and is the specimen most obviously corrupted.

The medical note is the foundational data of healthcare (Rosenbloom et al., 2010). Increasingly, it is just a copy of itself, as Chapter 1 already showed. A record that is half copied forward has a currency problem, because you cannot tell which findings are today's. It has a correctness problem, because a resolved condition carried forward is now wrong. And, it has a concordance problem, because the copied text and current lab results are drifting apart. 3 of the 5 dimensions, failing inside a single note.

The physician's reasoning is increasingly absent or buried in pages of generated text, and the next physician will never find it.

Recovery is an informatics problem and a clinical one. Template design, structured capture, natural language processing, speech recognition, ambient documentation, and the documentation integrity programs meant to hold all of it together are each part of the answer (Kreimeyer et al., 2017; Hodgson and Coiera, 2016).

THE TERMINOLOGY PROBLEM

Healthcare data depend on shared terminology and shared identifiers: a single definition of *this patient*, a single definition of *diabetes*, a single definition of *discharged home with hospice*. Master-data management and terminology services are the unglamorous infrastructure that holds those definitions in place. Where the infrastructure is healthy, data flow cleanly across organizations. Where it is not, the same patient appears in three different records, the same diagnosis is coded differently in each department, and the same outcome is calculated differently in each dashboard.

The treatment is the deliberate investment in master-patient-index management, in terminology services that map local concepts to standard vocabularies such as SNOMED CT, LOINC, RxNorm,

ICD-10-CM and ICD-10-PCS, and CPT, and in a governance process that prevents new local terms from entering without a mapping. These are called data standards, and since 2005 informaticists have been publishing why health data standards are required, how they are made, and what should be done (Hammond, 2005).

This is all clinical informatics work, and more than any other part of the data environment, it determines whether the rest of it can be trusted. Chapter 17 takes up the national dimension of the same problem.

WHY THE PATIENT RECORD EXISTS

Before counting the cost, it is worth remembering why the EHR industry was originally built, because most people working inside of it have never been told.

The EHR was not commissioned as a communication tool. It was not commissioned as a scheduling system, a messaging platform, a patient portal, or a place to keep notes. It has become all those things, and some of them it does well. *None of them was the reason.*

The record was commissioned as an instrument of measurement.

Read the statute carefully and the intent is not subtle. HITECH conditioned its payments on three things: 1) using a certified record in a meaningful manner, 2) exchanging health information to improve care, and 3) submitting clinical quality and other measures electronically. The three stages were named in plain language. Stage 1 was data capture and use in practice. Stage 2 was transmitting data to others electronically. Stage 3 was data analytics, specifically regulatory analysis.

That is a ladder, and every rung of it is made of data.

The vision underneath the statute was older and more ambitious. In 2007 the Institute of Medicine convened a roundtable and named what it wanted: *a learning health system*, in which science, informatics, incentives, and culture align for continuous improvement, with best practices embedded in the delivery process and new knowledge captured as a by-product of care (Olsen et al., 2007).

Every encounter would generate new information. That was the promise.

The healthcare industry was being asked to do what every other sophisticated industry had already done before. Aviation had been learning from its own mistakes for decades. So had manufacturing, logistics, banking, and retail. Each of them instrumenting the work, collecting the results, and using them to get better and make money. Healthcare, the largest sector of the American economy, was still running on paper.

So, the EHR was built to end that.

The list of things data were supposed to make possible is worth stating in full. *Quality* that could be measured rather than asserted. *Billing* substantiated by the record rather than argued after the fact. *Public health surveillance* in something close to real time. *Safety* signals detectable across populations rather than one sentinel event at a time. *Research* conducted on the ongoing care delivered rather than on the narrow slice of it that ends up in a research trial. *Payment* tied to something demonstrable. *Comparison* between institutions on terms both could accept. And, underneath all of it, an industry that could finally learn from its own work.

Decades of work and an enormous investment later, none of these have been achieved.

Look at what a nine-figure digital record sits on top of, in most American health systems. Paper forms. Fax machines. Scanned documents that arrive as pictures of text that are unsearchable, filed under a category no question will ever reach. Departments running shadow spreadsheets because the system cannot answer their basic questions. Manual abstraction staffed to produce the numbers the electronic measurement was supposed to produce.

The act of entering the data, the very thing the whole enterprise was built to accomplish, has become the defining complaint of the clinical workforce. That is the most worrisome finding of them all. Data entry is the largest single consumer of the working day and the most consistently measured driver of burnout and attrition (Sinsky et al., 2016; Melnick et al., 2020). The effort is a substantial share of the organization's administrative cost and explains why a physician who trained for a decade now spends more time typing than thinking.

We are paying physicians to produce data, at the highest labor rate in the building, and much of what they produce cannot be used.

THAT IS THE FAILURE UNDERNEATH EVERY OTHER FAILURE IN THIS CHAPTER.

Not that the record is hard to use, though it is. Not that it interrupts, though it does. The record was commissioned to make healthcare measurable and therefore improvable. It largely has not delivered and is consuming the workforce in the attempt.

The regulation that ordered EHR use carried no serious specification for the data. The interests that shaped it also had little incentive to write one. What arrived was adoption without a specification, measurement without a scale, and an industry instructed to capture everything without ever being told what any of it was for.

This is one of the most frustrating aspects of the field, and the thing a Diagnostician is best positioned to fix, because it is not a technology problem. The challenge is knowing what the data are for, designing the work so that those data are the natural by-products, and holding the data proposition accountable end to end.

Which brings us to what the failure costs.

WHAT CORRUPTED DATA COSTS

Data quality is usually argued on principle. It should be argued more on the invoice. So, here is the invoice.

Corrupted data do not present as corrupted data. They present months later, in disguise.

They arrive as a denied claim.

Premier's national survey of 280 hospitals puts the cost of contesting a single denial at $57.23, against a denial rate near 15%, with roughly 7 in 10 eventually overturned and paid. So, 7 in 10 of those claims were correct all along. Only their presentation had failed (Premier, 2025).

They arrive as a quality measure abstracted by hand.

Casalino and colleagues found American physician practices spending 785 hours per physician per year on quality reporting, a great deal of it extracting values from data the record already holds, done manually because the correct query was never built (Casalino et al., 2016).

They arrive as a case-mix index.

They report that the patients were healthier than they were. The organization is then underpaid for care it did not appear to deliver.

Not one of them looks like a data problem on the day it is discovered. The denial looks like a revenue cycle problem. The quality looks like a clinical problem. The case-mix index looks like a coding problem.

Three departments, three budgets, three sets of people working hard on three symptoms of one larger issue that none of them can see.

Every failure above has a price, and the price is not small.

A quality measure that cannot find the care that was delivered produces a payment adjustment the organization does not deserve. Incomplete or inconsistent coded bills produce denials that must be appealed, and every appeal is staffed by people whose entire working day exists because data elements go missing. Frequently.

Definitions that differ between departments produce the effort to reconcile, forever. An AI model that is a mismatch for local deployment produces clinical review, and in the worst case a reportable event.

Do you trust your numbers?

When hospitals are asked whether they *trust* their electronic quality measures, a substantial share say that they believe their manually abstracted numbers better reflect the care they delivered. Think about what that means. They are forced to run the required electronic measurement and then run a parallel measurement that they believe. That is two measurements where only one was purchased. The manual effort exists entirely because the digital effort is not trusted, and the people staffing it are doing work that would not exist in an organization whose data proposition was managed.

Chapter 3 established that administrative burden is the largest single category of avoidable cost in American healthcare, having no obvious treatment. Here is one specific remedy. Data that cannot be trusted create administrative work at every step downstream. The work compounds because each correction is manual and none of it fixes the source.

Repairing corrupted data is not a technical nicety. It is the key lever for understanding the largest cost categories in the industry, and it requires a Diagnostician who can trace a denial in November back to a build decision in March.

A PRACTICAL AUDIT DISCIPLINE

When consulted, I recommend a quarterly *data-lineage audit* of "high-stakes" metrics.

Pick a quality measure that drives a value-based-payment. Pick a financial metric that drives a board-level conversation. Pick an operational metric that drives a staffing decision. Pick an AI deployment that influences clinical care.

For each, begin with the question Kahn and colleagues put at the center of the framework. "*Fit for what use?*" Write down the intended use of the data before you assess anything, because a specimen adequate for one purpose may not be adequate for another.

Then walk the data capture end-to-end, from the bedside observation through capture, mapping, extraction, aggregation, and decision. Compare the dashboard number against an independent chart audit. Document the variance. Where the variance is large, examine the whole process.

This sounds onerous, and it is. *The cost is smaller than the return.*

There is another return on this investment that arrives more slowly. Once the staff expects that their work will be audited, they begin to enter critical data more carefully. The audit changes the culture faster than it changes the data.

BEDSIDE QUESTIONS: THE EXAMINATION OF YOUR DATA ENVIRONMENT

1. Which of the five dimensions of data quality has your organization ever formally tested a dashboard against, and can your leadership team name them?
2. For each of your most consequential metrics, have you written down what use the data are meant to measure?
3. Who owns the master-patient-index and the terminology services in your organization?
4. For your top five quality metrics, has the data lineage been audited end to end in the last 12 months?
5. Are your dashboard definitions stable across departments, or does each leader read a slightly different number?
6. Which of your clinical measurements are known to perform differently across patient subgroups, and who is responsible for knowing that?
7. Do you run manual chart abstraction in parallel with electronic measurement because you do not trust the electronic measurement? What does that parallel effort cost annually?

8. Have your AI deployments been locally validated against the actual patient population, with subgroup performance monitored over time?
9. Who is responsible for catching definitional drift between systems, and what is the escalation path when it is found?
10. Has your executive team sat with an analyst for an afternoon in the last year and walked a metric from definition to dashboard?
11. When the dashboard disagrees with the bedside, which one wins, and who decides?

Worrisome answers may mean the specimen data are contaminated, and the results are being reported anyway.

THE SPECIMEN IS CONTAMINATED, IN SIX RECOGNIZABLE patterns, each with a named cause, and a named treatment. A harder question arrives next. What if the treatment itself is the source of the harm? Some of the most dangerous medicines in the history of clinical practice were prescribed by confident physicians who had not been shown what they do to a patient.

CHAPTER 6

IATROGENIC HARM

WHEN THE TREATMENT CAUSES HARM

Every powerful therapy can harm a patient. The history of medicine is, in part, the history of recognizing that the treatments we prescribe with confidence sometimes hurt the patients we meant to help. Health IT is at exactly this point in its life cycle. The Diagnostician is the executive best built to recognize the harm, characterize it, and prescribe the change.

An 82-year-old woman is admitted for a serious infection, and the resident starts her on an antibiotic that must be dosed according to her weight, too little and the infection gets ahead, too much and her kidneys pay the price. The order set pulls her weight automatically from the chart. The available weight is three weeks old, recorded before a week of poor appetite from a stomach bug that took twelve pounds off of a frame that did not have twelve pounds to lose.

The dose that goes into her IV matches the number in the chart exactly. It does not match the woman in the bed. The number isn't mistyped or wrong. No warning was given or ignored. The system pulls the most recent weight on file, calculates correctly, and delivers precisely what the formula said to deliver. The nurse who hangs the bag does everything right.

When the lead physician investigates the injured kidney, they are looking for a mistake. What they find instead is flawless care, delivered by a system working exactly as it was built to, on a patient whose weight hasn't been recorded in three weeks.

IATROGENIC HARM IN HEALTHCARE

Aspirin, in the doses commonly prescribed in the early 20th century, hurt patients before clinical pharmacology caught up to the dose-response curve. Hormone replacement therapy was eventually shown to alter breast cancer and cardiovascular risk in ways that revised the standard of care. Spinal fusion, opioids, and once-popular bariatric procedures were all prescribed in good faith, by competent physicians, for plausible reasons, and each was eventually shown to harm patients in ways the earliest literature had not anticipated.

None of that story is bad medicine. It is the story of medical science learning, slowly, what its treatments do to a patient.

Iatrogenic harm is when the therapies given to treat the patient cause harm. It is common, and persists across every era of medicine.

Health IT is in the same phase of the same arc. The benefits, with the right configuration, are real. So are the harms with the wrong one. The Joint Commission, Office of the National Coordinator (ONC), and ECRI have called out that substantial risk (TJC, 2015; Sittig et al., 2014; ECRI Institute, 2024). The ONC has even published structured frameworks for assessing EHR-related risk.

The pattern is no longer novel, and it is well described.

This chapter addresses how the modern Diagnostician recognizes iatrogenic disease of this technology, treats it, and prevents the next case.

When the prescription is hurting the patient, the standard of care is to change the prescription.

In healthcare, we keep refilling it.

WHAT HEALTH IT SAFETY MEANS

Ask most executives whether their EHR is safe and you will get an answer about uptime, complaints, and whether the vendor has issued alerts. That answer covers roughly a third of the question.

Singh and Sittig, who have spent two decades building the measurement science for this, divide health IT safety into three domains (Singh and Sittig, 2016).

Domain I: Safe health IT. Is the technology itself sound? Free of defects, malfunctions, and design flaws that could injure a patient regardless of how carefully it is used.

Domain II: Using health IT safely. Is the technology being used appropriately, comprehensively, and in workflows that have not quietly become unsafe since it was installed?

Domain III: Using health IT to improve safety. Is the technology actively catching failures before they reach the patient?

Now place your organization inside of those three.

Domain I is largely the vendor's territory; it is contractually well defined, and most health systems monitor it adequately because a role is being paid to do that work. *Domain III* is where the value is, and almost no organization has reached it, because it requires the other two to be solved first.

Every pattern in this chapter lives in Domain II, "using health IT safely." And, in most American health systems, this has no owner.

This is a structural observation, and Sittig and Singh's own framework explains why it happens. Their sociotechnical model describes eight dimensions in which health IT operates: hardware and software, clinical content, the human-computer interface, people, workflow and communication, internal policies and culture, external rules and regulations, and system measurement and monitoring (Sittig and Singh, 2010). The dimensions are interdependent rather than sequential. A failure in one propagates through to the others.

That list does not easily map onto a hospital's organizational chart. *Hardware* belongs to IT. *Clinical content* belongs partly to physicians, pharmacy, nursing, and to service lines. *Workflow* belongs to operations. *Policy* belongs to compliance. *Regulation* belongs to legal. *Measurement* belongs to quality.

This overlap has no accountable owner: the job this book has been describing, and why the harms detailed below keep recurring in organizations where the job is seemingly going well.

TECHNOLOGY RELATED IATROGENIC DISEASE

Over two decades of clinical informatics literature has characterized five recurring patterns of harm associated with health IT. A 6th, drawn from my own practice rather than the published literature, rounds out that picture.

They are recognizable, preventable, and related to human factors rather than the technology.

Alert-fatigue-driven harm.

The decision support that should have intercepted a dangerous interaction was overridden with the rest of the *"boy who cried wolf" noise*, by a physician who long ago stopped reading the alerts. This pattern is documented from the early computerized provider order entry (CPOE) literature through the most recent meta-analyses, and it accounts for a meaningful share of preventable adverse drug events (Van der Sijs et al., 2006; Phansalkar et al., 2013; Co et al., 2020). The treatment is calibration. The harm is real, the treatment is available, and most organizations have not completed it.

Wrong-patient harm.

The order entered on one patient and intended for another, in an EHR without adequate identity verification, occurs at a rate of roughly 58 per 100,000 orders (Adelman et al., 2013). Chapter 1 discusses the therapy. A verification prompt at order entry reduces that rate by 16%. Organizations that implemented the intervention closed the gap. Organizations that did not are accumulating these errors.

Workflow-mismatched harm.

The EHR is configured in a way that does not match the clinical work it was meant to support, and the mismatch produces errors the bedside clinician would have caught had the workflow fit. The catalog begins with Koppel and colleagues' 2005 study and includes fragmented medication displays, default doses misaligned with the clinical context, pharmacy substitutions overriding prescriber intent, and interface designs that subtly invite the wrong action (Koppel et al., 2005; Slight et al., 2016). The treatment is workflow co-design with the people who do the work, governed as a permanent capability rather than a project.

Missed-result harm.

The abnormal lab, the unread imaging, the critical value that surfaces in a notification system no physician is reliably reading (Singh et al., 2009; Murphy et al., 2017). In the early studies, roughly 7% of abnormal results lacked timely follow-up, with electronic-notification design identified as a contributing factor. The treatment is closed-loop result management: explicit ownership of every critical-result pathway, structured handoff when the ordering physician is unavailable, and audit of the closure rate as a primary safety metric.

Algorithmic-error harm.

The AI model deployed without local validation or monitoring under-identifies one patient subgroup, over-identifies another, or recommends a plan that does not match the local standard of care. When Obermeyer and colleagues examined a widely used population health algorithm, the arithmetic was correct, the proxy was corrupt, and the model failed patient subgroups (Obermeyer et al., 2019). Daneshjou and colleagues found the same failure in a different modality, where dermatology models performed worse on darker skin than the curated benchmarks suggested (Daneshjou et al., 2022).

Physician-pacification harm.

This last pattern has no settled name in the literature. I call it *physician pacification syndrome*: the condition in which the physician trusts that the technology has done what it was supposed to do and stops checking. The vital sign that was measured was recorded. The message that was sent was received. The therapy that was ordered was delivered.

The assumption is both understandable and dangerous. Health IT has introduced outcomes with no analog in paper charting or human-to-human work, because clinicians falsely assumed the system was automatically validating or monitoring the data.

The governance response is identical to the five before it. A physician-leader who knows where the system fails will warn the others and is accountable for catching it. (See Figure 6.1.)

IATROGENIC HARM · WHEN HEALTH IT HURTS

Iatrogenic Impact of Technology

Six recurring patterns of preventable harm from ungoverned health IT.

PATTERN 1

Alert-Fatigue Harm

The alert meant to intercept danger gets overridden along with the noise.

PATTERN 2

Wrong-Patient Harm

An order entered on one patient, intended for another — about 58 per 100,000.

PATTERN 3

Workflow-Mismatch Harm

EHR configuration produces errors the bedside clinician would have caught.

PATTERN 4

Missed-Result Harm

Abnormal labs sit in queues no clinician reliably reads — about 7%.

PATTERN 5

Algorithmic Harm

A model deployed without local validation, and the bias goes systemic.

PATTERN 6

Physician Pacification Syndrome

The clinician trusts that the system already did what it was supposed to do.

Named in this book.

Five patterns are documented in the literature. All six are preventable with governance.

FIGURE 6.1

Iatrogenic harm from health IT — five documented in the literature, the sixth named here.

THE MISSED DIAGNOSIS

I have walked into organizations where several of these patterns were active at once and they had not been recognized as a common syndrome.

The IT department was making clinical build decisions on its own, responding to individual tickets rather than to any pattern. The clinical staff was struggling with workflows that had not been designed for the work they were doing. The executive team was managing a workforce in open revolt and watching the dashboard metrics fall.

A piece of this is owned by all. The patients and clinicians knew it. The numbers showed it. The IT help desk tickets pointed straight at it.

The person whose role was to identify all of the symptoms, diagnose them, and treat them, was missing.

The same absence of clinical leadership was producing the deficient note, the alert fatigue, the frustrating visit, the quality decline, the payment reduction, and the physician who stopped checking.

The absence of that synthesis is not a minor gap in the organizational chart. It is how patients get hurt.

MEASURING THE WRONG VITAL SIGNS

There is a measurement problem underneath "harm," explaining why these patterns persist in organizations that genuinely care about safety.

Almost every health system counts adverse events *after* they occur. That is necessary work, but it is not sufficient. A safety program built entirely on retrospective event reporting will only find the harm that was noticed and reported. The physician who overrode an alert did not experience an event. The critical result that was not communicated generated no report. The decision support that was supposed to flag an issue produced nothing at all, and the harm was hidden in the care that never happened.

The Health IT Safety framework asks for something different: proactive, ongoing measurement in all three listed safety domains rather than a body count in one (Singh and Sittig, 2016). The instruments already exist. The Joint Commission named health IT an institutional hazard in Sentinel Event Alert 54 and called for formal safety processes. The ONC published the SAFER Guides precisely so organizations could self-assess in a structured way. ECRI has ranked health-technology hazards among its Top 10 Patient Safety Concerns every year since 2012 (TJC, 2015; ONC, 2025; ECRI Institute, 2024).

None of that information is proprietary. None of it is expensive. Most organizations have never systematically used any of it.

The obstacle is not the available instruments. It is that the role of assuring the safe use of the clinical technology, remains absent.

WHAT HARM COSTS

The financial and reputational consequences of unchecked iatrogenic harm, in 2026, are measurable and substantial.

Beyond the cost to human life already described, there is also a number that can be assigned.

Quality-linked payment adjustments under value-based purchasing swing reimbursement by between *2% and 9%* of total Medicare revenue, and a meaningful share of those swings are determined by how reliably the decision support, the clinical content, and the workflows produce the structured trail required by the quality assessments. Hospital-Acquired Condition penalties, a 1% reduction in Medicare payment, are imposed annually on the worst-performing quartile, with the determination flowing from the same EHR-captured data this book has examined. Readmissions penalties continue. Information-blocking disincentives for physicians risk their exclusion from the Shared Savings Program (ONC, 2024).

There is a third consequence, and it arrives years later through the legal system.

The Candello national malpractice database, drawn from insurers across the country, found serious documentation issues in 1 out of every 5 professional liability cases. Having issues with bad documentation in a liability case more than doubles the odds of an indemnity payment (Candello, 2024).

The claim is not that better records prevent malpractice, though they may. The claim is that the record is evidence, that the evidence is often poor, and that the difference between a defensible case and a paid claim is decided by the forces that created that record.

A direct chain runs from decision-support mismanagement, through alert fatigue, through preventable safety events, through reportable-event volume, through quality scores, through the payment adjustment, to the operating margin.

Every link in that chain is informatics mediated. Every link sits inside the reach of a Diagnostician who has been given the authority to act on it. And, every link has been documented to respond, predictably and measurably, to the governance this book describes.

There is another ledger underneath all of this.

EACH OF THESE HARMS GENERATES ITS OWN *ADMINISTRATIVE WORK*: THE event review, the appeal, the corrective action plan, the re-audit, the regulatory response, and the litigation. Administrative burden is the largest single category of avoidable cost. Iatrogenic technology is one of the machines that manufacture it.

BEDSIDE QUESTIONS: THE EXAMINATION OF YOUR IATROGENIC RISK

1. Which of the three domains of health IT safety does your organization actively measure, and can they name all three?
2. Has every reportable patient-safety event in the last 12 months been examined for an informatics root cause?
3. What is your current alert-override rate, and what is the trend?
4. Has your organization implemented a patient-verification forcing function at order entry?
5. Is your closed-loop critical-result management workflow documented, audited, and owned by a named person?
6. Has your organization ever completed a structured self-assessment using the SAFER Guides? When, and what happened to the findings?
7. Are your AI deployments locally validated, with subgroup performance monitored over time?
8. Are your value-based-payment adjustments improving year over year? If not, why not?
9. When care goes wrong because information failed to reach the person who needed it, does your safety program record that as a communication failure, or does it record it as a clinical one?
10. Who in your C-suite is accountable for iatrogenic technology harm as a category? If there is no answer, the category is unmonitored.

Unanswered and badly answered questions mean that your EHR is causing harm and is not being addressed.

. . .

THE IATROGENIC PATTERNS ARE NAMED, AND THE RESPONSE TO each is available and documented. What remains is whether the organization has built the structure where that response becomes a decision. Complex care requires an interdisciplinary team working together. Chapter 7 examines the institution built precisely for that.

CHAPTER 7
THE MULTIDISCIPLINARY CARE CONFERENCE
TEAM MEDICINE

The multidisciplinary care conference is the structure by which a complex patient gets one coherent treatment plan instead of several siloed ones. Healthcare organizations need the same mechanism, and the same Diagnostician to convene it.

THE TEAM IS ELEVEN CLINICIANS STRONG BY THE TIME IT REACHES ROOM 412, and the patient has learned to recognize the sound of it before they knock. The patient is three weeks into a hospitalization that started as pancreatic cancer and has now become an obstruction their body cannot resolve on its own. For three of those weeks, the patient and family have asked and answered the same questions to a new team member every morning.

Today the whole group stands at the foot of the bed at once. Surgery reports what the imaging ruled out overnight. Nutrition raises the fact that the patient has not tolerated anything by mouth in four days and asks whether it is time to feed them a different way. Palliative care says the pain regimen is undertreating them at night and overtreating them by noon, and the pharmacist, standing at the back, catches that the treatment nutrition wants to add would blunt the one pain medication working. The disagreement happens in front of the patient instead of in four separate notes they will never read.

The family member in the chair by the window asks, "What is happening?" For the first time in three weeks, one person answers, and every clinician in the room nods. The team moves on to room 413. Nothing about the cancer changed that morning. What changed is that eleven people left the room having agreed and said the same thing.

THE MANY BECOME ONE

A patient on a complex medical service in a large academic hospital is, on any given day, the responsibility of multiple consulting services, several non-physician disciplines, and one physician-leader of the case. The cardiologist has been asked about the heart failure. The nephrologist about the abnormal kidney function. The surgeon about the unhealing wound. Pharmacy about the list of medicines. Care management about the discharge plan. Each consultant answers a focused question, leaves a focused note, and returns to their other patients. None of them is expected to hold the whole picture.

The whole picture is assembled by the team conference. The team conference is led by the lead physician.

In a well-run service, the conference can happen daily. The team gathers, sometimes in person, sometimes by phone, sometimes by messaging, and discusses the patient as a group. Each consultant says what they think, in roughly two minutes, in language the rest of the team can follow. The lead physician synthesizes. Conflicts are surfaced and resolved. Plans are adjusted in real time. The team leaves the conference with one plan rather than five, and the next steps of care are written from that plan.

THE CONFERENCE IS NOT CEREMONY

Multidisciplinary team meetings have been studied most rigorously in oncology, where the structure is most formalized, and the findings are consistent.

Lee and colleagues followed 85 consecutive cases presented at gynecologic oncology tumor board conferences over 16 months and measured what the conference accomplished. And, it showed an impact. The treatment plan changed in 27.1% of cases, with 10.6% of those being major changes, meaning a substantial treatment was added, removed, or changed. The diagnosis itself changed in 9.4% of cases. And, 91.8% of the conference's treatment decisions were implemented (Lee et al., 2017).

A physician would rate this intervention as a success.

Better than 1 in 4 of the patients left the conference with a different plan than the one they arrived with, 1 in 10 of those with a materially different treatment, and better than 9 in 10 of those changed plans were carried out. No individual produced that change. The group did.

Then ask the harder question. "*Does the conference change what happens to the patient?*"

Kesson and colleagues had an unusual opportunity to find out. In 1995, multidisciplinary team work was introduced across regional areas of Scotland, while neighboring areas continued as before. The authors followed 13,722 women with invasive breast cancer diagnosed between 1990 and 2000 (Kesson et al., 2012).

Before the change, the intervention area was the *worse* place to have breast cancer, mortality running 11% higher than the comparison area. After the change, it was the better place, with mortality 18% lower than the comparison area.

The authors report one further finding that belongs in conversations about governance. Multidisciplinary work not only improved survival, *it also reduced the variation in survival between hospitals.*

Nothing else about those hospitals changed in that interval. No new drug arrived. No new operation was invented. The intervention was that a group of clinicians began meeting to produce one plan instead of several, and women lived longer.

The conference is not ceremonial activity. It is the intervention. Stated plainly, governance is coordination.

What Kesson measured is what coordination looks like at the bedside. A defined chair holds it. Every relevant discipline is present. The data are read together instead of apart. One plan binds until the next meeting revisits it.

Governance is the identical mechanism, run on the organization instead of the patient. Not an analogy to it. The same discipline, practiced on a different body.

The chief financial officer (CFO) has a focused view. The chief information officer (CIO) has a focused view. The chief medical officer (CMO), the chief nursing officer (CNO), the chief operating officer (COO), the chief quality officer, the general counsel, and the chief human resources officer each have a focused view. None of them individually is expected to hold the whole picture.

The whole picture is held by the team conference. And, the team conference must be held by a lead who is fluent in all organizational languages, clinical, technical, operational, financial, and regulatory, and who is empowered to synthesize.

What follows is how that conference works, why so many organizations have built a poor substitute for it, and why the Diagnostician is uniquely qualified to convene and chair the synthesis.

MULTIDISCIPLINARY COMMITTEES

A well-run multidisciplinary clinical care conference has five characteristics, each of which can be reproduced in a healthcare organization.

It is chaired by a leader with responsibility for the whole patient. The chair is the physician of record, holding the diagnosis and writing the plan.

It is attended by representatives of every relevant discipline, prepared to speak briefly, in shared language, about the slice they see. The cardiologist says, in two minutes, what they think about this patient and what they recommend. Each executive arrives prepared to contribute their slice in shared language.

It is grounded in data. The labs, the notes, and the consults are visible to all. The conference is where the data are read together. Organizationally: the metrics, the trends, and the audits are visible, and the conference is where they are interpreted together.

It is short, frequent, and disciplined. Short and frequent forces the discipline of decision-making. The patient cannot wait for next month's meeting.

It concludes with a single written plan which is reviewed at the next meeting against measured progress. In organizational governance the plan is captured, the owners identified, the expected actions defined, and success criteria agreed upon. (See Figure 7.1.)

THE MULTIDISCIPLINARY CARE CONFERENCE

Good Healthcare Governance

Five habits that turn a meeting into a synthesis – clinical and operational, at once.

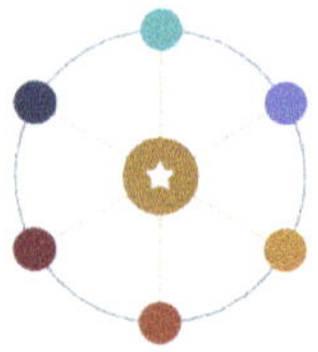

		CLINICAL	OPERATIONAL
1	**Chaired by a Synthesis Leader**	Owns the diagnosis, writes the plan.	Accountable for the final call, across every domain.
2	**All Disciplines Present**	Every voice heard, in plain shared language.	Executives contribute clearly, not in jargon.
3	**Grounded in Actual Data**	Labs, notes, and consults reviewed together.	Metrics and audits read together, never summarized.
4	**Short, Frequent & Disciplined**	The patient can't wait for the next cycle.	Frequency forces decisions, prevents ceremony.
5	**One Written Plan**	Signed by the chair, reviewed next time.	One page: decisions, owners, criteria, date.

The five characteristics that distinguish a synthesis from a meeting.

FIGURE 7.1

What good governance looks like, mapped clinical to operational.

WHEN THE CONFERENCE GOES RIGHT

A well-run organizational conference is remarkable for how undramatic it is.

The patient is presented: the recent margin trend, the recent quality dashboard, the recent denial pattern, the recent safety event, the recent technology decision. Each consultant speaks briefly. Conflicts are surfaced. Decisions are made. The plan is signed off. The next meeting reviews progress. The pace is steady.

What is remarkable is what does not happen.

The IT-led project that would have been quietly approved without clinical review does not get approved. The vendor pitch that arrived on the CFO's desk before the team considered the nursing implications gets paused. The decision-support rule that would have been activated in the next upgrade gets a risk assessment first. The denial pattern that would have been treated as a billing problem gets traced back to a gap in the clinical content and fixed at the source. The performance of an AI tool that would have been deployed is measured before deployment rather than after.

Each of these five decisions, paused before they brought conflict and failure, carries a cost the organization never pays. A functioning synthesis is one of the few mechanisms that prevents the poorly coordinated work rather than perpetuating it.

WHEN THE CONFERENCE GOES WRONG

Several ways that conferences fail are worth naming, because they are common.

The rubber-stamping session. The meeting happens. The slides are presented. The minutes are recorded. No decisions are made or changed. The synthesis is not happening. The meeting produces minutes and not plans.

The captured conference. One of the specialists has become the agenda-setter, and the synthesis bends toward that viewpoint. Others attend dutifully, but the work is no longer cross-disciplinary.

The under-empowered conference. The right people are in the room, the right discussion is happening, but the committee charter cannot commit the organization to the resulting plan. The synthesis arrives at a recommendation, the recommendation is sent to the *actual* decision-maker who was not in the room, and the questions are revisited from scratch. The pace of decision-making slows. The team disengages.

The under-instrumented conference. The data on which the synthesis depends are not reliably available, so the discussion runs on impression and anecdote rather than measurement. Decisions get made on the wrong information, and the synthesis drifts away from the underlying reality.

The treatment in each case is recognizable. Charter the conference with the right authority. Place the right leader in the chair. See that the data are reliable, visible, and current. Hold the meeting with discipline. Produce a plan, sign it, distribute it, execute it, and review it.

Then do it again next week.

BEDSIDE QUESTIONS: THE EXAMINATION OF YOUR CROSS-TEAM SYNTHESIS

1. Does your organization have a chartered cross-disciplinary and cross-executive synthesis function, chaired by a lead physician with explicit authority?
2. When the CFO, CIO, COO, CMO, and CNO disagree on a question that is simultaneously clinical, technical, operational, and financial, who decides? Is the answer a name, or a process?

3. How often does your Diagnostician (CMIO or equivalent) actually preside over a decision that crosses multiple disciplines? If the answer is never, your cross-disciplinary and cross-executive synthesis function is not chartered.
4. Are your governance meetings producing single-page plans, or producing minutes?
5. Are your discipline-specific committees nested under a cross-disciplinary and cross-executive synthesis function, or operating in parallel?
6. In the last year, how many proposed decisions were paused or reversed because the synthesis caught something no single committee would have seen? If none, the synthesis is not working.
7. Has any senior leader in your organization sat in the CMIO's seat for a day in the last year?
8. If your CMIO resigned tomorrow, would the cross-disciplinary and cross-executive synthesis "Function" survive, or would the committees revert to parallel operation?

If you cannot answer these satisfactorily, they describe a synthesis problem. Parallel processes, siloed decisions, and no single owner driving the plan.

THE CROSS-DISCIPLINARY AND CROSS-EXECUTIVE SYNTHESIS FUNCTION HAS BEEN DESCRIBED. The conference is convened, the plan is written down, the team is in the room. We now need to examine the chair of the meeting.

PART III

THE MAKING OF A DIAGNOSTICIAN

The executives who hold the modern clinical informatics budget, hire its leaders, and set its mission have been making these decisions for two decades, and largely do not understand what clinical informatics is, what it requires, what it should do, or what it should cost. You cannot select for something you cannot describe. You cannot resource a role you do not understand.

This is not a verdict on every executive suite. I have sat across the table from CFOs who read a balance sheet the way I read an EKG, and from physician CEOs who ask sharper data questions than most CMIOs I have trained. Organizations already led by one look different in the outcomes ahead, and the chapters that follow show where. But two decades of national spending without national improvement is not explained by the exception. It is explained by the rule.

Part III examines the role, the industry, and the return on the investment. It addresses the financial questions for this field that have never been answered before.

What is all of this costing, and what would it cost to fix?

The answer begins with six questions, and they take about ten minutes.

CHAPTER 8

INVESTING IN THE DIAGNOSTICIAN

NAME, FORM, AND FUNCTION

A PHYSICIAN WHO IS ASKED TO CONSULT ON A COMPLEX PATIENT WITHOUT time, compensation, and authority, will eventually decline to consult. The same applies to the physician-informaticists, doing informatics work for decades as a side duty.

A 68-YEAR-OLD WOMAN IS RECOVERING ON THE MEDICAL FLOOR AFTER A bout of pneumonia when she mentions that her right knee has been aching for weeks and is now making it hard to get out of bed. The hospitalist rounding on her that morning knows the knee needs an evaluation, but the hospital's only orthopedic surgeon is out for the week. Rather than wait, the hospitalist pages cardiology, the only consultant with time on the schedule that day.

The cardiologist arrives, checks pulses and swelling, and writes a note. It is a careful note, written by a genuinely skilled physician, offering the wrong specialty's opinion on the problem.

The mismatch is not physician incompetence; it is a clinical need without the optimal specialist.

THE ROLE AS PLACEBO

Roughly 7 in 10 American acute care hospitals now employ a person with the title of CMIO, or some variation of it (HIMSS, 2019). And, every healthcare symptom documented so far persists.

There are only two available conclusions. Either the role does not work, or many of those with titles are not in the right role or at the right dose for what we are discussing. The evidence leans toward the second explanation.

Consider what a *placebo* is. It is an inactive preparation administered in the form of a real treatment. It has the right container, the right label, and the right ritual. It produces hope, without real biological change. Deployed at scale it yields exactly the pattern we observe: low cost of production, widespread use, an occasional victory, but no actual impact on disease progression.

An executive role deployed without the ingredients for success is also a *placebo*.

It has the right container. There is a title on the organizational chart, a name in the directory, a seat at some meetings, and a line in the annual report describing the organization's commitment to clinical informatics. It has the right rituals like committees, roadmaps, and quarterly updates. What it does not have is the active ingredient to address the disease.

American healthcare has spent the last two to three decades preparing the *placebo Diagnostician*. The defining property of a good placebo is that the people taking and giving it cannot tell the difference. Unfortunately, neither can most hospital boards. The active ingredients have never been defined.

NAMING THE TREATMENT

The most common Name being used for healthcare's executive for informatics therapy is *chief medical information officer (CMIO)*. Occasionally, other names are given. The success of a therapy is never in its Name; it is in its active ingredients. And, inconsistent naming is only where the confusion starts.

Every other executive title in a hospital tells you exactly what the person controls. The chief financial officer controls the money. The chief nursing officer controls nursing. The chief information officer controls the technology. You know the job because you know the Name, and who reports to it.

Nothing is supposed to report to just a Name. And, in most American hospitals, that is the honest and most common description of this one.

I have held this title in academic centers, community hospitals, children's hospitals, and integrated delivery systems. In some of them I had a budget, a team, and the lead physician role. In others, I only had a business card. The work was the same in all cases. The results were not remotely the same, because they were most correlated to the organization's commitment to the role.

So, the Name is not the subject of this chapter. The Name is free. Any organization can print one this afternoon, and in today's healthcare market, around 70% already have. Of note, that means 3 in 10 hospitals have so little commitment to the function that they haven't even assigned a free Name.

MANUFACTURING THE RIGHT FORM

Before a curative treatment can begin, the medications need to be manufactured. The right medication containing active ingredients, at the right dose, with the right timing, and absorbable by the patient.

In healthcare, the Form of the therapy is what the organization has to commit to: six ingredients, and each of them are decisions leaders in your building can make.

Education and Preparation. Screening whether the person is trained for this: a strong graduate degree in clinical informatics, not nothing, and not one of the many placebos contaminating the name, maybe a fellowship or a clinical informatics board certification, always weighted experience. This is not a role for a physician with some interest and a little time.

Funding. There is a line in the budget that they decide how to spend, not a loan from another colleague's resources.

Positioning. They are in the room, with a vote, where the decisions get made.

Empowering. Their vote garners true administrative weight. They can *stop* something. Authority is not only the right to be consulted, but the ability to halt an errant decision.

Staffing. They have people who report to them, are chosen by them, and whose efforts are directed by them, not only a borrowed resource when they aren't busy elsewhere.

Employing. This is their full-time job, not a title stapled to a busy clinical schedule, and not a stipend.

That is the Form. It is not complicated or expensive, simply rare.

You can test your own organization against it in about 10 minutes, and you do not need to know anything about technology to do it. Ask yourself whether your organization employs a role with these six ingredients.

Is the person leading your clinical technology strategy trained for it, and can you name the credential? Do they control a budget, and what is the number? Were they in the room at the last major clinical, administrative, or technology decision, and did they speak before the vote? Can they stop the organization from doing something? Do people report to them? Is this their whole job? *Six questions, six points.*

The ability to say "no" is the only one that reveals actual decision-making authority. Ask the room to name the last important decision that person stopped. In most organizations what happens next is a pause, and then the explanation that it does not really work that way here. That pause is your finding. Write down a zero and keep going.

Five or six means the Name and Form are present for Functional success. Three or four means the Name exists in some Form but the Function is constrained by the organization's commitment to the role. One or two means the Name is present and the Form and Function are not. Zero means that you may not even have the *Name*.

DEPLOYING THE RIGHT FUNCTION

The Name is free. The Form is **who** you have committed to. The Function is **what** outcomes become possible once you have committed. If you learn only one thing from reading this book, please learn this: the **Function** is essential for healthcare healing, and the numbers that prove this are coming soon. Whatever you decide to name it, seek it.

Here is the plainest description of the Function I can give. The Diagnostician needs to read the whole chart. Not just the technology chart, the entire organization's chart: the quality scores, the denial reports, the turnover data, the safety events, the override rates, and the documentation audits, at the same time, looking for the single upstream cause that explains several of them at once. *That is the job. Everything else follows from it.*

They must be a **clinician** because every downstream data and function of a healthcare organization starts at the clinical encounter. A dose, alert, physical exam layout, documentation template, and AI setting are clinical decisions, currently being made in your building, and rarely being made by a clinician with the training and license. The reason they must be trained in **informatics** is that reading that chart requires knowing how the data got there, every step. A physician who cannot tell you why the dashboard disagrees with the bedside situation cannot fix the dashboard. And, the reason they must sit in the **C-suite** is that the fixes take time, cost money, and cross departments, and they will require the authority to say "no."

Clinical judgment, technical fluency, workflow engineering, and executive standing. Miss any one and the Function does not run. Chapter 9 examines each and what breaks when it is missing.

WHY HALF A COMMITMENT FAILS

Here is the pattern I have watched more times than any other. An organization decides, for some reason, that it needs this *Name*. It cannot yet justify the whole *role*, so it only does part of it: names a respected physician, gives them a stipend and some administrative time, and asks them to represent clinical interests in the technology environment. The effort involved is done in good faith, the decision looks prudent, and it can be worse than doing nothing, because the Name checkbox has been filled. Doing nothing at least leaves the problematic gap visible. The half commitment makes it seem to disappear. The organization now believes the *Function* is covered. The board hears that clinical informatics leadership is in place. The question stops being asked.

Even when they have capability, the person cannot do the important work this book describes, because what they should be doing was never defined. Another common error is thinking of the role as a part-time effort away from the clinic. They will have no time for the important informatics work, because they are busy trying to prove to their colleagues that they are still a doctor. The Function done properly consumes a full week of scientific and cognitively demanding work; the two roles do not fit.

Two or three years later the physician burns out and goes back to practice. The role is rechartered. The next willing physician is found. Nothing has changed except that a good doctor has been used up.

There is a second cruelty in the half-committed arrangement, and it works right into the argument.

THE CONFIGURATION THAT STOPS THE ROLE FROM WORKING ALSO STOPS others from finding out that it is not working.

A person who has never been asked to hold a contrary position against the chief financial officer is never revealed to be unable to. A person briefed on the quality scores, without having to interpret and respond to them, is never found to lack the ability. A person never asked to repair a broken clinical workflow or dashboard does not reveal that they cannot. So, the organization concludes it has a Diagnostician, because they have the Name. The appointee concludes they are doing the job, because they still have the Name, and have never been asked to do anything complicated enough to threaten it. The illness and symptoms of the organization continue, and not one person in the arrangement is lying.

That is not a personnel problem. The role and what it was supposed to produce were never written down. If you understood even a fraction of the function's return on investment, you would never attempt it halfway. You would throw everything into it. The rest of this chapter is an attempt to make that value plain enough that halfway measures never get tried again.

WHAT THE RIGHT SETUP LOOKS LIKE

The following are descriptions of the conditions that produce success.

- Reporting to the chief executive,
- Working partnerships with the chief medical officer and chief information officer,
- Leading the multidisciplinary governance cadence,
- The performance review against defined outcomes,
- The decision authority that separates them from their peers,
- The permanent chair in the executive committee, at strategic planning, at capital allocation, and in the board reporting cycle,
- Employment rather than rented participation,
- Defined deliverables,

- Dedicated time,
- Executive compensation,
- Executive accountability,
- A talented team of informaticists, data scientists, and process improvement experts that reports through this role rather than around it, the way finance staff belong to the chief financial officer.

That last one is the part organizations resist, and it is also the part that costs the least. Which brings us to the money.

THE DIAGNOSTICIAN'S RETURN ON INVESTMENT

What follows is the value. A return on investment (ROI) is a fundamental financial metric used to evaluate the efficiency or profitability of an investment. It is the reasonable expectation of money that will be made or returned by dividing the net profit of an investment by its fiscal cost:

ROI = [(\$Value - \$Cost)] ÷ (\$Cost)

An official one has never been offered for the Diagnostician or clinical informatics, though plenty of people have measured around its edges. I will price it in the way I would work up any patient: the size of the problem, then the cost of the treatment, then the return. Every number has an identified source and reason for its selection. Because a direct measure does not exist, we build the value proposition here.

The fact that this has never been done, despite the vast healthcare expenditure and illness we have been describing, supports the book's premise: a condition that is never diagnosed is never treated, and a treatment never given cannot be shown to fail. Pulling together numbers and thoughts from different studies into one hypothesis is a challenge, and this book is the first to bring this important set of numbers together. The arithmetic is deliberately simple, because I am not an economist.

Every number will be converted to:

$dollars/physician/year

This produces a common unit for comparison, useful for three reasons: it is reliably published and updated, it is stable, and it is a number every executive already knows for their own organization. Multiplying it by your physician count gives you your own number.

If you do not agree with the numbers, prove them wrong. Every challenge is welcome, because this is meant to be the conversation starter; at least, someone will finally be paying attention.

Let us start the calculation.

READING THE PATIENT'S BILL

Before pricing a patient's therapy, it is worth knowing how sick they are, what they are already spending, and on what. In healthcare, the numbers are staggering. The United States spends $5.3 trillion a year on healthcare (CMS, 2025c); no other wealthy nation is close, and the per-person figure runs roughly twice what our peers pay. In 2024, the Association of American Medical Colleges (AAMC) counted 1,032,365 physicians in the U.S. (AAMC, 2024). Divide by the national price tag and it comes to $5,114,470 a year for every practicing physician, about *$5.1 million/physician.*

Healthcare money is not all the same.

Split the bill in two, because the two halves are not the same kind of money.

First, waste, money an organization does not intentionally spend.

There are eight categories of measured healthcare waste, added for the first time here, that cost the U.S. $867.5 billion a year, roughly $840,304 per physician of what we will call *waste exposure*. It is the most studied money in American healthcare, and it is the most important in this chapter (Mello et al., 2010; Sinsky et al., 2016; Han et al., 2019; Shrank et al., 2019; Forrester, 2026; NSI, 2026).

This leaves $4.41 trillion for ordinary healthcare operations. Nurses, beds, supply chains, drug spend, imaging, capital. This converts to $4,274,166 per physician. This is the real cost of doing business, intentionally spent, with far less literature on what of it is recoverable.

The Physics of Healthcare Revenue and Recovery.

Money is not one flowing liquid like classical economics treats it; a dollar of revenue does not carry the same weight. A dollar of new revenue is subject to margin, the inescapable cost of producing it, in supplies, personnel, and the administration of the enterprise. In the U.S., margin averages around 3%, meaning the cost of delivering care consumes 97% of everything healthcare brings in.

Loss recovery runs on a different law. A dollar lost is a dollar the organization had already earned after all the spend that produced it, meaning it took $33 at that same 3% margin to have $1 to lose in the first place. A recovered dollar carries none of that drag; it drops straight to net profit, rescued from the void rather than squeezed from new volume.

The Diagnostician's informatics Function performs across both laws. It repairs the technology, data, and workflows that clinical care and operations run on, and the repair can land on either side of the ledger. An inefficient or lossy process, once repaired, returns loss recovery at $1 recovered for $1 gained; a lowered cost of care behaves the same way, $1 lower cost for $1 gained. Improving the revenue engine itself, increasing output and volume, benefits the organization only at the margin, $1 of new revenue for $0.03 gained.

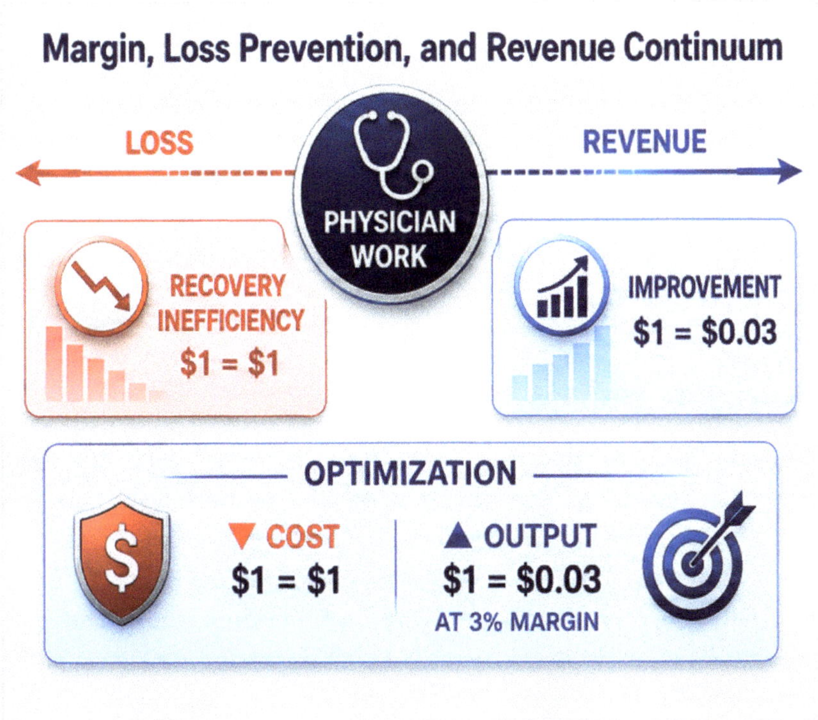

What follows is where, on America's balance sheet, that optimization has the most room to work.

THE MONEY WE SPEND ON PURPOSE

Starting with the larger half, $4.41 trillion or $4,274,166 per physician per year, is what it costs to run American healthcare. Chapter 3 took that apart line by line and the shape has not changed. Hospital care is 31% of it. Physician and clinical services are 21% of it. Pharmaceuticals are 13 to 14%. Long-term care, insurance overhead, and public health make up roughly 30% more. Devices, supplies, and equipment are 2% to 4%. Nurses, beds, supply chains, drug spend, imaging, and capital are the rest.

The questions a Diagnostician must ask are whether better operational and clinical workflows, technology configuration, and data that are delivered faster, to the right person, at the ideal moment make an enterprise reduce its losses or improve revenue capture. In a high-performing organization, the questions are whether good clinical workflows, care at standard, and outcomes can be improved. Those are different questions from recovery, and the answers are elusive.

I went looking for that answer, and found mostly an absence of a coherent question.

A Literature Without a Diagnosis

Three studies have tried to measure the operational impact of well-configured technology, and what they mostly discovered was how bad the underlying data, site variation, and methods were. Evidence on healthcare IT, operational dashboards and command centers among them, shows highly variable outcomes, ranging from real cost savings and added bed capacity to increased wait times and new data problems. Johns Hopkins improved efficiency; other sites saw negligible or negative results, meaning benefit depends on implementation quality rather than the technology itself (Coiera et al., 2025; Kane et al., 2019; Mebrahtu et al., 2023). Broad economic projections for healthcare IT are considered unreliable, hampered by flawed methods, undocumented costs, and data that cannot be compared across studies; most studies even lack randomized controls, and one analysis found the outcomes varied too widely across sites to generalize, suggesting standardized health-economics checklists for future work (Low et al., 2013; White et al., 2023; Coiera et al., 2025).

In 2008, the Congressional Budget Office rejected a RAND Corporation projection of $81 billion in annual health IT savings, finding it relied on unrealistic assumptions of immediate universal adoption, ignored implementation costs, and extrapolated gains from a handful of high-performing sites (Congressional Budget Office, 2008). No federal body has revisited the question since, across the whole period in which adoption has become nearly universal. The National Academy of Medicine published a standard model for assessing institutional return in 2014 (Adler-Milstein et al., 2014); I found no evidence of its use.

The most defensible reading of this literature is that local configuration and data usage varied wildly by site and produced real, if inconsistent, operational effects that have never been generalized, and their causes never established.

There are three ways to read a literature that behaves like that. The *first* is that informatics has no operational effect and the local findings are noise. The *second* is that hospital operations are too intertwined to attribute a saving to any single cause. This is the reading the field has been living on for two decades. The *third* explanation that I find most credible is that the operational research methods used were simply insufficient.

An Educated Guess

So, what is the Diagnostician's opportunity for ROI in this category? The answer is not published and no one appears to be looking.

What I would say is that it is a vast unexplored territory. In my opinion, it would make no sense that designing optimal operational and clinical workflows, technology configuration, and data, delivered at optimal speed, at the optimal time and in the optimal place, would not substantially improve the opportunity. And, we have never tried.

Maybe the best way to envision this is that healthcare is a fusion of sciences that will continue to develop. Medical science, administration science, AI, drugs, devices, and procedures are all on a lifetime improvement pathway, and the learning health system is an eternal aspiration, not a destination (IOM, 2012). Informatics needs to evolve with all of them. The alternative to the Diagnostician is hoping that it all works out OK.

What cannot continue is the current pathway. American healthcare has been so staggered by the rapid pace of change since HITECH that asking questions about the "optimum state" seems like science fiction compared to the "barely functional current state" this book has been presenting.

If forced to estimate the size of this opportunity, I would put it between 1% ($44 billion) and 5% ($221 billion) of the $4.4 trillion, above the savings from basic EHR implementations that Rhoades measured and below the 5-10% national modeling from Sahni, Cutler, and colleagues (Sahni et al., 2023), and below the per-admission savings Kazley and colleagues found in hospitals with an advanced record (Kazley et al., 2014).

WE DID ALL OF THIS WORK AND I AM NOT GOING TO COUNT A DOLLAR OF IT. The temptation is there because I believe it, but the hesitancy comes from my skepticism of every number in the field. Nothing has meaningfully advanced since RAND published its 2005 projection and the Congressional Budget Office dismantled it three years later; Kellermann and Jones revisited RAND's projection in 2013 and said plainly that the technology had been capable and it was the organizational redesign that had been missing (Kellermann and Jones, 2013).

We can now examine recovery of loss, where the data exist.

THE MONEY HEALTHCARE WASTES

Now the smaller half. Waste is a loaded word, and we must narrow its meaning. I mean money that leaves an organization without buying any care, in ways measurable enough that researchers have put national figures on them. One of the most important questions of them all has never been meaningfully published. "*How much is all of this costing us?*" And, it reveals a large opportunity for return that informatics investment can deliver.

Measuring Healthcare

American healthcare is one of the most heavily measured enterprises on earth. We know what a hip replacement costs, what a hospital bed-day costs, what a nurse costs to replace, and what it costs to appeal a denied claim. *These numbers have never been put on one page*, not because they are secret, but because every consultant has examined only one organ system, one operational exposure. Then they left a note and never read the whole chart. So here are the notes, together.

Shrank and colleagues did the foundational work. Their 2019 review built six domains of waste in American healthcare and priced each one: administrative complexity, pricing failure, failure of care delivery, overtreatment or low-value care, fraud and abuse, and failure of care coordination (Shrank et al., 2019). I will use four of those six and decline two. Pricing failure, the inflated cost of goods, is declined because informatics does not set prices; fraud and abuse is declined because informatics does not monitor crimes. Those two are large and loud categories in waste taxonomies; leaving them out costs this argument several hundred billion dollars before it starts. I do not concede that there is nothing that can be offered for their improvement, but I will stick with the four numbers I can defend.

Four more sources sit alongside Shrank. Sinsky and colleagues priced the clinical day with a stopwatch. Han and colleagues priced burnout and attrition. Mello and colleagues priced the liability system. Forrester priced the technology that we briefly examined earlier (Sinsky et al., 2016; Han et al., 2019; Mello et al., 2010; Forrester, 2026). They are all published, and none of these has ever been added together before, because the people who study waste do not study burnout, and no job description has ever included the arithmetic.

These categories represent money that a Diagnostician, a well-governed clinical enterprise and EHR, and a designed informatics strategy could reach.

Four rules were used to produce the total below. Take the low end of every published range. Exclude anything informatics cannot obviously reach, pricing, fraud, anything requiring a change in law rather than a change in workflow. Where two sources price overlapping ground, keep the smaller figure and drop the overlap, which is why several numbers below run lower than the papers they come from.

And, where a category deserves an estimate rather than a zero, provide one and name it.

The result is eight categories and $867.5 billion a year: paperwork that wastes time, the half-day that doesn't count, losing our clinicians, care we did not need to give, harm and what follows it, money we earned and did not collect, technology we pay for and do not use, and a spillover category of other measured exposures. All eight of them rest on named published measurement; the money we didn't collect is priced at the conventional share, up to 5% of net patient revenue, that revenue cycle work is measured against, $35 billion, I will use a 2% conservative value, easy for math, and under the conservative estimate of 2.68% in the literature (Rotenstein et al., 2026; Gartner, 2016). Divided by the same 1,032,365 physicians, the $867.5 billion is ***$840,304 per physician per year.***

Exposure and Opportunity.

Exposure is the size of a category in which loss is happening, not a claim that all of it is recoverable. It tells you how much money is at stake while the decision about funding this Function gets made. The true return depends on whether an exposure is fully curable or only healable. I have left out billions that were even mildly questionable, taken the low end of every category below, and set a conservative opportunity target of 2% for each. I leave the final numbers to the economists, but the Diagnostician return is significant even at this estimate.

Paperwork That Wastes Time

This is the administrative work created by many payers with different rules, permissions sought before treating, claims argued after treating, and quality reported in formats that were never automated. They cost $265.6 billion a year, and become the largest of Shrank's six domains and the only one lacking a documented, evidence-based fix. Physicians and their staff spend 13 hours a week on prior authorization alone. Contesting a denied claim costs $57.23 and roughly 7 in 10 denials were correct but lacking data the first time. Practices spend 785 hours per physician per year reporting quality measures.

Most of this is just data moving badly between systems that already hold them. A claim denied for a missing element sitting inside the chart is not a payment dispute, it is a failed handoff inside the organization's own systems; a quality measure abstracted by hand from a discrete field is a query never automated. Priced at the flat 2% rate we will claim as opportunity, this category is: $5.3 billion (Shrank et al., 2019; AMA, 2025; Premier, 2025; Casalino et al., 2016).

The Half-Day That Doesn't Count

Sinsky and colleagues followed physicians with a stopwatch and found 49.2% of their day going to the EHR and desk work, against 27% in direct contact with patients; priced against physician compensation, that is roughly $180 billion a year of clinical time spent on data entry. This doesn't even address the opportunity for additional revenue generated. Some of it is recoverable through structure alone, fewer clicks to place an order, a template that pulls forward what the chart already knows instead of asking again. The larger and newer mechanism is ambient AI: Rotenstein and colleagues tracked 8,581 ambulatory physicians across 5 academic health systems and found that adopting an AI scribe was associated with 16 fewer minutes of documentation time and 13.4 fewer minutes of total EHR time for every 8 scheduled patient hours, among physicians who used it on half their visits or more. Instead of assuming that AI magically cures provider burnout overnight, the data shows that technology may offer a small slice of relief. Real efficiency requires changing fundamental operational workflows and reducing systemic clinical documentation requirements.

Priced at the flat 2% rate we will claim as opportunity, this category is: $3.6 billion. This is also the category with the most directly measured informatics effect in this chapter, since Sinsky's stopwatch and Rotenstein's trial both measured time rather than inferring it; 2%

likely understates this one. The figure also counts only physicians. Nurses document too, and time-and-motion studies put their share of a shift between a seventh and a third of it. The KLAS Arch Collaborative surveyed 80,147 acute care nurses across 179 organizations and found 79% report meaningful time lost to unproductive charting; where organizations optimized it, Mercy recovered 32 minutes of charting per nurse per day, Wooster Community Hospital recovered more than 15,000 nursing hours a year, and Children's Hospital of Philadelphia booked over $21 million in savings from one initiative. Advanced practice clinicians, pharmacists, and therapists are not counted at all. The true size of this category is larger than the $180 billion, and probably substantially so (Sinsky et al., 2016; Rotenstein et al., 2026; KLAS Research, 2025).

Losing Our Clinicians

A new focus across the healthcare industry lately is clinician **wellness**. Wellness represents everything that can burn out a hard-working clinician, and there are many. In almost every survey and study of the subject, the EHR gets named as a top issue. Clinicians leave because of attrition and burnout, and replacing them or renting their substitutes costs about $30 billion a year: nurse turnover runs $19.5 billion, agency and travel premiums about $5.4 billion, and physician burnout, counted only through turnover and reduced hours, another $4.6 billion, $7,600 per employed physician per year (Han et al., 2019). Replacing one bedside nurse costs $60,090, and the gap between a travel nurse and an employed one runs $66,081 a year, per position (NSI, 2026). Replacing a physician costs far more: a Stanford case study priced recruitment at $250,000 to nearly $1 million per physician depending on specialty and academic rank, found physicians reporting burnout more than twice as likely to leave within 2 years, and put the 2-year institutional loss from 58 departures at between $15.5 million and $55.5 million (Hamidi et al., 2018).

The KLAS Arch Collaborative surveys more than 121,000 clinicians; 43% of burned-out nurses name the EHR as a cause, and nurses who left their organizations had scored the EHR 5.1 points lower than nurses who stayed (KLAS Research, 2026a). Organizations that ran serious optimization work gained 6.2 points on average, more than the entire gap between leavers and stayers. The mechanism is the same for physicians. Shanafelt and colleagues surveyed 6,375 physicians in active practice and found lower satisfaction with clerical time and higher burnout among those using an electronic record and computerized order entry (Shanafelt et al., 2016). Melnick put a slope on it: every one-point gain on the usability scale is associated with 3% lower odds of burnout, and the median American EHR usability scores 45.9 out of 100 (Melnick et al., 2020). Sieja and colleagues put an eleven-person optimization team into clinics for two weeks each and moved measured burnout from 39% to 34% in the same population (Sieja et al., 2019). This result and KLAS's organizational scores prove that the slope moves when the underlying cause is addressed.

Priced at the flat 2% rate we will claim as opportunity, this category is: $600 million. National average nurse turnover is 17.6%, the best decile of hospitals runs 11.5%, and $600 million is a fraction of what closing even part of that gap is worth. Advanced practice clinicians, who deliver a large and growing share of American care inside the same record under the same configuration, are not counted at all; there is no defensible national figure yet for what their turnover costs.

Care We Did Not Need to Give

This is a category that contains a few differing concepts. All of the figures are at the low end of their ranges. The first represents the tests that were repeated because the first result could not be found, and treatment delivered that the evidence does not support. This costs about $103 billion per year. An EHR record that cannot find a prior result is a data and configuration issue, not a limitation of medicine; studies of health information exchange find repeat emergency imaging falling by 44% to 67%, when the prior study is visible

(Lammers et al., 2014). Shrank's failure-of-care-coordination consists of fragmented inpatient and outpatient care that leads to unnecessary admissions and readmissions. This costs about $27.2 billion per year. The entire EHR data model was formed around improving this fragmentation. Overtreatment and low-value care runs $75.7 billion per year. These are unnecessary tests, prescriptions, and care that are unlikely to benefit patients given potential harms, high costs, or better alternatives. And, these impacts are complicated by the model of reimbursement of the environment (Shrank et al., 2019). On the overtreatment side, what appears as the default in an order set is often what gets ordered, and those choices are being made today. Most of this waste is not paid the way a la carte care is. Medicare's Inpatient Prospective Payment System pays a hospital one fixed amount per admission, by Diagnosis-Related Group (DRG), regardless of how many extra drugs, tests, or days that admission consumes; ordering something unnecessary inside a DRG-paid stay generates no additional revenue, only additional cost (MedPAC, 2021). The largest piece of this category, $44.1 billion of the $75.7 billion, is overuse of end-of-life care, concentrated in exactly this kind of inpatient stay. Decision support has already shown it can stop this specific pattern: an alert against one unnecessary repeat test cut those orders 21% at one health system (Levick et al., 2013), and a duplicate-order block saved another $183,586 over two years (Procop et al., 2014). Priced at the flat 2% rate we will claim as opportunity, this category is: $2.1 billion.

Harm and What Follows

These are the patients hurt by the care they received, and everything that follows the harm. It cost about $112 billion a year in direct terms. Shrank's failure-of-care-delivery category is $102.4 billion (Shrank et al., 2019), and the medical liability system adds roughly $10 billion in indemnity payments and administration (Mello et al., 2010), setting aside the cost of defensive medicine. The Candello national malpractice database finds documentation failures in 20% of liability cases, and when present more than double the odds of an indemnity payment (Candello, 2024).

Order entry with decision support has cut serious medication errors by half or better in published trials; Kaushal's team at Brigham attributed the largest single component of a $28.5 million return to exactly this mechanism (Kaushal et al., 2006), and better documentation reduces legal exposure directly.

Priced at the flat 2% rate we will claim as opportunity, this category is: $2.2 billion. Mello and colleagues put the cost of defensive medicine at another $45.6 billion of the $55.6 billion total cost of the liability system, more than four times the indemnity and administrative figure counted above (Mello et al., 2010). I have left all of it out, because I cannot prove that better build and documentation reduce a defensive posture. Although, the fact that the presence of defensive medicine can be measured also suggests it also can be reached. The total in this chapter would be materially larger if I had counted it.

Money We Earned and Did Not Collect

This is care that was delivered and paid at a lower rate than it should have been, because the record did not capture how sick the patient was. Under-coding in healthcare is endemic and costs roughly $35 billion a year across documentation integrity gaps and charge capture leakage, using the low end of the conventional up to 5% of net patient revenue. Two different leaks share the line. Documentation integrity is care that was billed without the record showing how sick the patient was; charge capture is care that was delivered and never billed at all. An implant opened to the wrong account, a high-cost supply pulled and never linked, a bedside medication never charged, and charges posted inconsistently across interfaces. Correct payment is a documentation, workflow, and interface problem that informatics should own, including problem-list governance, helpful prompts built into the workflow rather than a billing query weeks later, and diagnosis capture built into the note instead of bolted beside it. Charge capture is duller and easier, every leak is a broken handoff between two systems the organization already owns, and there must be accountability for their handoff. Priced at the flat 2% rate we will claim as opportunity, this category is: $700 million, likely the highest-yield category in the ledger, possibly 15-50% instead of 2%.

Technology We Pay For and Do Not Use

These are the clinical applications that go unopened, licenses that go uncanceled, interfaces that are never retired, and add up to a substantial volume of legacy technology. Chapter 3 put American provider technology spending near $69 billion a year, about 1.3% of national health expenditure; what share of it is dead weight is never published, and every organization that has looked has found some. Some governing body must hold the clinical inventory list and have the authority to switch things off. This is simply a matter of ongoing governance and accountability rather than clinical change. Priced at the flat 2% rate we will claim as opportunity, this category is: $1.4 billion, it is probably the easiest money in this chapter to collect.

Other Measured Financial Exposures

This is a spillover category that catches what does not fit neatly inside the other seven. I limited it here to two components with named national figures, but open it to additions when they have been quantified. The American Hospital Association's Manatt Health study puts regulatory compliance burden at $38.6 billion a year across hospitals and health systems (AHA/Manatt Health, 2017); the Agency for Healthcare Research and Quality puts the aggregate cost of potentially preventable inpatient stays at $34.3 billion a year (McDermott and Jiang, 2020), together more than $72.9 billion, administrative staff time on one side and foregone hospitalizations on the other. The American Hospital Association and Manatt Health's own breakdown of the $38.6 billion shows meaningful use and quality reporting, structured data capture, EHR certification, automated measure extraction, at 19.3% of the total, and billing and coverage verification, the same claims and documentation work priced above, at another 21.6%, better than a third of the figure before counting anything else. McDermott and Jiang's $34.3 billion is admissions for ambulatory care sensitive conditions, diabetes, chronic obstructive pulmonary disease (COPD), asthma, hypertension, heart failure, pneumonia, urinary tract infection, the textbook target of population health informatics: a registry flagging a patient overdue for an A1c, an alert on an uncontrolled blood pressure, a coordination platform keeping a COPD exacerbation out of the emergency department. Priced at the flat 2% rate we will claim as opportunity, this category is: $1.5 billion.

Putting It All Together

Add those eight categories and the country is exposed to roughly **$867.5 billion a year** in losses that good clinical informatics can potentially reach, $840,304 per physician. Every figure comes from a published source, and every range is taken at its low end. Waste, simplified, runs about **$840,000 per physician per year.**

THE TOLL ON OUTCOME

There is a second ledger, and it does not convert to a monetary number well. 795,000 Americans are permanently disabled or killed every year by a misdiagnosis. More than 1 in 4 physicians report experiencing a delayed prior authorization that led to a hospitalization, a permanent impairment, or a death. Between 44,000 and 98,000 patients are killed by preventable hospital error on the conservative national estimate, and credible work puts the real figure considerably higher. Those are the same categories priced above, counted in human impact instead of dollars.

I will not put a monetary price on that column. It deserves more respect than a dollar amount. Every figure in this chapter is only a proxy for it, the part of this disease that happens to be measurable.

The money matters, but it is not the part that matters most, because the core calling is doing no harm.

IF THE ARITHMETIC ABOVE IS WHAT MOVES YOUR BOARD, USE IT. I would prefer they were moved by something else.

THE DOSE OF INFORMATICS TREATMENT

Against that $5.3 trillion, what does the country already spend on the healthcare IT instrument and Diagnostician?

On the technology, $69 billion of provider technology budget, $66,837 per physician per year, about 1.3% (Forrester, 2026; AAMC, 2024). $0.013 of every health dollar buys the record that every other cent runs through.

On the Diagnostician Function itself, the discipline of pointing that technology at a clinical priority, remains untracked. There is no national figure. What we know is that 81% of physician informatics leaders do not have a budget of their own: 19% hold their own leadership budget, 40% are funded inside the IT budget, 8% are funded somewhere else, and 33% have no budget at all (WittKieffer and AMDIS, 2025).

Beneath that sits something structural. Worksheet A of the Medicare cost report defines every general service cost center an American hospital is required to report against: capital, employee benefits, administrative and general, plant, housekeeping, dietary, nursing administration, central services, pharmacy, medical records. There is no cost center for information technology, information systems, health IT, or clinical informatics; computer hardware and software costs are named explicitly in the instructions and directed into administrative and general (CMS-2552-10). The national financial reporting system that governs every hospital in this country cannot see this function even in principle. And, the definitive national study of the health IT workforce, which counts 161,160 staff, does not separate clinical informatics from general information technology and reports no cost data at all (Hersh et al., 2018).

There is no benchmark because there is nothing to survey; no attempt has been made to look for it. This supports the argument running through seven chapters. An organization that had genuinely constituted this Function would have a line for it. Most do not, and the federal cost report does not offer them one. Health IT, the category Chapter 3 identified as the instrument through which the largest category of avoidable cost is amplified or reduced, is filed inside that largest category, which is why so little of this spending appears on a line called health IT. What follows is the first price I am aware of published for this function. I am not competing with a benchmark. I am supplying the one that does not exist.

PRICING THE DIAGNOSTICIAN'S THERAPY

I have spent years listening to people explain that the business case is unclear, the cost is too high, and the role is not a priority. *The real situation is*: we are losing $840 million a year in a 1,000-provider system, at the low-end estimate, and the person positioned to reduce some of it costs $440,000. The exposure runs more than $1,900 to every $1 of salary. That is the opportunity.

I will price the budget for a properly built informatics program at a system with 1,000 physicians from other people's published numbers rather than my own. Sieja and colleagues published the full economics of one: a team of eleven, one physician informaticist, one nurse informaticist, one project manager, four EHR trainers, four EHR analysts. It costs $1.2 million a year, including the physician at about $280,000 with benefits, and reaches about 500 physicians a year, approximately $2,400 spent per physician touched (Sieja et al., 2019). National survey data describe the same object from the other direction: WittKieffer and AMDIS surveyed more than 160 physician informatics executives in 2025, and at organizations under $3 billion in revenue, 73% of these leaders have 10 or fewer direct reports (WittKieffer and AMDIS, 2025). The unit, then, is not complicated: one Diagnostician, and one team of eleven for every 500 physicians.

Price the Diagnostician the same way. Buckhead FMV read the Form 990 filings of 73 nonprofit health systems and found average CMIO compensation of $436,287 in 2024, 83% of these executives earning a base salary above $350,000, the largest single band at $401,000 to $450,000, 78% eligible for a bonus, and 69% still seeing patients (Buckhead FMV, 2024; WittKieffer and AMDIS, 2025). Call it $440,000 fully loaded, and both sources support it.

For 1,000 physicians, then: two teams and one executive. In the unit this chapter uses, the Diagnostician is $440,000 divided by a thousand physicians, $440 each; the two teams are $2,400,000 divided by a thousand, $2,400 each. The program costs $2,840 per physician per year. Multiply that reference value by your own physician count and you have your own number.

Now the question that changes the conversation: how much of that is new money? Look at the eleven again. Four EHR analysts, four trainers, a project manager, nine people your organization already employs, sitting in information technology, in training, and in whatever your version of a project management office is called. That leaves two. Many systems do have a nurse informaticist somewhere under nursing, almost none with protected time for this work; almost none have a physician informaticist at all, because protected time for a physician to do this work is the most common thing that goes unfunded. I count those two as new.

Sieja discloses the physician informaticist at $280,000, leaving $920,000 for the other ten, $92,000 each. 9 of 10 already on the payroll is $828,000 a team already being spent. The two new positions are $372,000 a team. Two teams plus the Diagnostician are $1,184,000 of new money against $1,656,000 already committed; **$1,184 new spend and $1,656 already committed per physician.** Nine of the eleven already work for you. One executive, two to four FTE clinicians given protected time to do work they already know is needed, and a reporting-line change for people already on your payroll: that is the whole ask, which costs 700 times less than the opportunity.

MANAGEMENT VERSUS A CURE

A physician makes this distinction constantly: some illnesses can only be managed, others can be cured. Nursing turnover can be lowered, but not stopped, because people move, retire, and change their minds; burnout can be lowered; and diagnostic error can be lowered. Most of medicine works that way.

But some losses exist only because they are left ungoverned; when they finally do get governed, they don't just shrink, they stop. A claim denied for information already sitting in the chart. A quality measure miscalculated by hand from a field the record already holds. A duplicate order inside a single instance of a single system. A license for software unused for 2 years. None of these have a champion; an unused application has no backers. They survive because no Diagnostician has ever been handed both the inventory and the authority.

About a quarter of the return comes from categories like these, and that quarter behaves differently from the rest. Addressed correctly, it ceases to exist entirely.

THE DIAGNOSTICIAN'S CONSERVATIVE RETURN

So, what is the ROI? Eight categories of measured waste, but one of them, Care We Did Not Need to Give, is not one thing. Shrank and colleagues, the paper behind this whole ledger, split it into two when they built it. Failure of Care Coordination, $27.2 billion, readmissions and avoidable complications from poor handoffs between providers. Overtreatment or Low-Value Care, $75.7 billion, medications, tests, and procedures that should never have been ordered. Both are billed and reimbursed. Only one behaves like the rest of this chapter once corrected.

I used 2% as the recovery rate across our loss domains, taking the low end of every published range. 2% of $840,304 is $16,806 per physician per year, $16.8 million for a system with a thousand physicians, $17.4 billion nationally. Against a program costing $2,840 per physician, that is a 5.9 to 1 return.

If 2% is wrong, change it. At 1%, the return is $8,403 per physician and 3 to 1. At 3%, it is $25,209 and 8.9 to 1. The number is almost certainly higher.

I tested that instead of asserting it. A category-by-category exercise against this chapter's own figures independently lands at 2.45%. Crediting only the two categories with real, published evidence behind them, Rotenstein and colleagues' measured minutes saved by ambient documentation tools (Rotenstein et al., 2026) and Gartner's finding that a mature software-governance program cuts spend 30% in its first year (Gartner, 2016), discounted hard for a program still finding its feet, a literature-specific review lands at 2.68%. Both are higher than the flat 2% this chapter actually uses. I am reporting the lower number anyway for simplification and consistency.

Then the number for the skeptics, the one that ends the arguments about risk. Not what will this deliver, but how little could it deliver and still be worth doing. The program costs $2,840 against an exposure of $840,304; to pay for itself, it must recover only $0.34 (full program cost) of every $100 this organization is already losing.

One last way to see the return. The average U.S. physician creates around $2,380,000 of annual revenue (Merritt Hawkins, 2019), $71,400 after a 3% margin adjustment. For a system with a thousand physicians, the Diagnostician brings in $16.8 million of loss recovery, over a 230 to 1 comparison.

From the outset, informatics has been treated as a cost center. A cost center has no return. This one returns 5.9 times what it costs at the most conservative rate. (See Figure 8.1.)

THE HIERARCHY OF THE NUMBER

Why $0.34 Changes Everything

What one physician costs the system, and what oversight recovers, per year.

$5,114,470

TOTAL U.S. HEALTH SPENDING, PER PHYSICIAN, PER YEAR

$5.28 trillion ÷ 1,032,365 active physicians (CMS, 2025c; AAMC, 2024)

$4,274,166

Operations — nurses, beds, supplies, drugs, capital. No claim made.

$840,304

Measured waste, eight cited categories. The target.

v

PER PHYSICIAN, PER YEAR	
EXPOSURE *the eight priced categories*	**$840,304**
CONSERVATIVE RETURN *2% of the exposure*	**$16,806**
PROGRAM COST *1 Diagnostician + 2 teams of eleven*	**$2,840**
NEW MONEY REQUIRED *nine of eleven already on staff*	**$1,184**

THE BAR THIS FUNCTION MUST CLEAR

$0.34	**5.9:1**	**700:1**
recovered per $100 of exposure	conservative return to program cost	financial exposure vs. the new-money fee

Scales linearly. Multiply every line by your physician count.

FIGURE 8.1 · Pricing Informatics Using Data

What one physician costs the system, and what oversight recovers, per year.

HOW THE PATIENT HEALS

No competent physician promises a cure this size quickly. Healthcare has a chronic disease which has been neglected for a long time, and it will take time and effort to respond.

If the return is this large, why has it gone unidentified? I do not fully know. I have been making this argument for decades with plenty of evidence. The EHR legislation, informatics scientists, healthcare researchers, insurance payers, and finance leaders have all shown enthusiasm for good electronic data, and have been waiting for the opportunity. Most have missed it. Ask healthcare's best and brightest what the underlying deliverable of the EHR and its data collection was supposed to be, and the answers, or the quiet, will surprise you. Some of it is a lack of awareness of the informatics function, discussed already in this book. Some of it is a data-science answer handed to a nation that has fewer than 15% of its population holding degrees in science, technology, engineering, and mathematics (STEM) fields, a trillion-dollar industry that ran for centuries on paper, and untethered expectations doing the rest.

The better answer is that this data-led approach is still novel to many, and despite significant illness healthcare is wary of novelty; it is running a care plan that should have been abandoned years ago. Make no mistake: this is hard work, sustained by a fatigued workforce, over a long time, baked into ongoing ledgers. It is not a quick fix, but it is the only fix, and full maturity is a decade of work. Kaushal's program at Brigham did not turn a net profit until year six (Kaushal et al., 2006); a faster timeline than that should not be promised. That is not a weakness in the argument, it is the reason the argument has never been made. A return that takes six years to show up is invisible to an executive team that turns over in four, and that invisibility is exactly how an ROI ratio this large has gone unexamined for over seventeen years.

THE FINAL VERDICT

Let me return to stating the argument the way a physician does.

There is a disease. Seven chapters have documented it, and it isn't in dispute anywhere. The disease costs at least $840,000 per physician per year, and that sum had never previously been added up. The treatment costs about $1,184 per physician per year in new money, because most of the team already works for you. To pay for itself, the treatment must recover only $0.34 (full program cost), or $0.14 (new money) out of every $100 of financial exposure, the symptoms. The most conservative reading of the published literature says it recovers multiple times that.

Year one pays for itself before clinicians are even asked to change how they practice.

I have made this argument to boards for two decades using adjectives. Adjectives lose a budget battle every time. This is the first version with arithmetic in it. The cost of the Function stopped being the interesting question the moment the financial exposure turned out to be 700 times larger than the fee. This is not a cost center, it returns what a hospital would otherwise have to generate through over 230 additional physicians, using the chapter's most conservative case.

Treatment is indicated. This patient needs a Diagnostician.

Which leaves the only question the arithmetic cannot answer, and it takes the next chapters. *Who is qualified to be one?*

BEDSIDE QUESTIONS: THE EXAMINATION OF YOUR INFORMATICS CAPACITY

1. Run the six questions on your own organization, asking your chief executive rather than your informatics leader. What is the score, and which conditions failed?

2. Name the last thing your clinical technology leader stopped. If the room falls silent, you have your answer to the fourth question.
3. Count the people in your organization already doing informatics work under some other leader. The analysts, the trainers, the documentation specialists, the report writers. What would it cost to add a qualified physician above them, and what are you paying for them now?
4. What did you spend last year on denied claims that were eventually paid, on nurse turnover, on agency labor, on prior authorization staffing, and on quality reporting? All five numbers exist in systems you already own. Have they been put on one page?
5. What is your total technology spend, and what fraction of it goes to the clinicians who decide what the technology does?
6. Are your physician informaticists paid for protected time, or doing the work as volunteers?
7. What financial impact can you attribute to your informatics program over the last 12 months? If you cannot answer, it is not being measured, and an unmeasured program cannot be defended at budget time.
8. Which of your losses are curable and which are only reducible? If they have never been sorted, start there. The curable ones produce evidence. The reducible ones produce argument.
9. If your informatics leader left tomorrow, who takes the role?

Three problems are hiding in these answers at once. A synthesis problem, an engineering problem, and a data problem. There is likely an undervaluation issue shared by many.

. . .

THE DISEASE HAS A PRICE AND THE TREATMENT HAS ONE TOO, AND now they are on the same page. Every organization is already writing the clinical technology prescriptions, every day, at scale, and in most buildings those people are not Diagnosticians. Chapter 9 examines who they are, if they are qualified, and why these questions have gone unasked.

CHAPTER 9
CLINICAL TECHNOLOGY DECISIONS
WHO SHOULD MAKE THEM

The non-physician leader who steps in to write a prescription because no physician is available is making the best decision they can with the wrong training. This is, in essence, what happens every day when clinical technology decisions are made outside of clinical informatics governance.

A patient on a busy hospital service is in pain. The bedside nurse has paged the lead physician. The page has not been returned. The pain is escalating.

The administrator on duty, an executive with 30 years of healthcare leadership behind them, makes an unfortunate decision in the moment. They pull an available pain protocol, identify a medication, and ask the nurse to administer it.

Every physician reading that paragraph has flinched.

In this example, the administrator's desire to help was admirable. The administrator's authority, in that moment, was not. Pain management, even though on a protocol, is a clinical decision. It belongs to the lead physician of the case. The administrator stepping in, however well-intentioned, is operating outside their training, outside their credential, and outside the scope of authority any reasonable hospital would assign to the role.

We do not allow this to happen at the bedside. We have entire systems, entire cultures, and entire credentialing infrastructures built for no other purpose than to produce that flinch, in the right person, at the right moment, before the wrong decision is made.

Now walk down the corridor to the room where the physician's order set, documentation template, and clinical workflows are being designed. There is no flinch in that room.

THE ROOM WHERE THEY DO NOT FLINCH

There is a project plan, a deadline, and an analyst with a configuration screen open, making a clinical decision that will be executed 40,000 times a month by people who will never meet them. The IT director is asked to help refine the decision support rule. The analyst is asked to set up the medication display logic. The chief information officer is asked to approve the artificial intelligence deployment. The compliance officer is asked to interpret the clinical regulation. The vendor's implementation team is asked to help configure the order set.

Each of those are clinical decisions.

That room is not acting in bad faith and has never been formally addressed. The administrator in the pain example was at least caught by a credentialing system built to catch exactly that mismatch. The analyst configuring the order set is given a title, a project plan, and a deadline.

And, here is what separates the two rooms.

At the bedside, we would know the consequences within the hour. In the configuration room, the consequence arrives 18 months later, distributed across the turnover rate, the denial rate, the burnout survey, the quality and safety report, and the contract labor line, in quantities small enough that no single one of them is ever traced back to the afternoon some analyst without clinical training answered a clinical question.

The pattern does not persist because the organization is careless. It persists because, in most healthcare organizations, the alternative has never been built. The analyst configures the order set because it is on a checklist and the screen is in front of them. The compliance officer interprets the clinical regulation because the deadline is in 2 weeks and the physician with the training is in clinic seeing patients.

Those are reasonable responses to an unreasonable situation, and the unreasonable situation is the absence of the Diagnostician who should have been making the decision.

This chapter is about who that person must be, why the qualification is so rare, and why the gap has gone unnoticed.

CREDENTIALING THE PRESCRIBER

Chapter 8 named the competencies the informatics Function requires and now we will address them in order. Clinical judgment. Technical fluency. Workflow engineering. Standing. They are not a wish list. They are the ways a clinical informatics process fails and each failure looks completely different from the outside.

They are also, together, the function. Not the certificate and not the title. This is what a Diagnostician must be able to do before letting them fill out the prescription pad.

Clinical Judgment and What Happens Without It

Most of the upstream data issues in this book are created by clinical input wearing technical clothing. The default dose of a medication that can injure the kidney is a clinical decision, and so is which order sets use it, which safety alert fires and on what criteria, what it asks the physician to do and when it is suppressed, whether a documentation template asks for a given finding in a given format, whether an AI model is safe to deploy on this patient population in this way, and whether an alert needs a hard stop or only a soft nudge. Getting it wrong in either direction causes harm. All of it presents as technical configuration settings of the EHR. All of it is clinical decision-making.

Without clinical judgment in the chair, the decisions still get made, by whoever has the screen open, has the time, and has a deadline.

The signature of this failure is an environment whose build is technically complete and clinically wrong. The interfaces are stable, the uptime enviable. The alerts fire reliably, and physicians override them between 49% and 96% of the time (Van der Sijs et al., 2006). The order sets are beautifully laid out and reflect a protocol that the organization has not used since 2019. The dashboard flashes a number and measures the wrong denominator. Nothing is technically broken. Everything is wrong.

Technical Fluency and What Happens Without It

The mirror failure is a physician in the chair who cannot read the machine. A physician who cannot explain why the sepsis dashboard disagrees with the bedside situation cannot fix the sepsis dashboard. They can convene a meeting about it. They can escalate it. They cannot fix it. The diagnosis lives in a mapping table, an interface specification, a numeric calculation, a value in a data set that was written long ago and has never been read since. It depends on how the data were captured, in what format, by whom, and what happened to it on the way to the dashboard.

This does not require the ability to write code, though that helps. It requires the ability to ask the questions the non-technical workforce misses: where does this number come from, which patients are in the denominator and which were excluded and by whom, what does this field mean in the original system, and does it mean the same thing on the dashboard.

The signature of this failure is a physician who is respected, articulate, and permanently one layer above the actual problem. They speak beautifully in the governance meeting. The build does not change. I have watched gifted physicians occupy this position. They were never told that the job required a second discipline, and no pathway existed to give it to them.

Workflow Engineering and What Happens Without It

This is the competency almost never named, and I would build most of the training around it. Healthcare technology projects predominantly fail due to unpredictable, real-world organizational and regulatory complexities rather than technical defects (Greenhalgh et al., 2017). Clinical judgment tells you what should happen. Technical fluency tells you what the system is doing. Neither tells you what the work looks like at two in the morning on a short-staffed unit, and that is a third discipline entirely. It is standing at the workstation, in the operating room, at the nursing station, in the clinic, long enough that people stop performing for you and go back to working. Measuring how long a task takes, every step of it, and how quickly the critical result safely reaches the right person. Mapping everything that touches the work. Then redesigning what you saw so the technology makes the clinical work faster, easier, or more accurate, instead of interrupting or degrading it.

This is systems engineering and process improvement. It is not a clinical skill. Medical training does not teach it, and neither does informatics training in most programs.

The signature of this failure is unmistakable: people working harder, not better, for all the effort.

Koppel and colleagues studied barcode medication administration (BCMA) at 5 hospitals, shadowing nurses, sitting in staff meetings, pulling the override logs. They found 15 distinct types of workarounds and 31 distinct causes (Koppel et al., 2008): patient identification barcodes affixed to computer carts, to the scanners themselves, to doorjambs, to nurses' belt rings; several patients' medications pre-scanned and carried together on a cart; barcodes crinkled, smudged, torn, missing, or covered by another label; scanners that malfunctioned; wristbands chewed, soaked, or absent; medications never barcoded; failing batteries; wireless that came and went; and emergencies, during which none of it mattered. Nurses overrode the system's alerts for 4.2% of patients and 10.3% of medications charted. The authors concluded that shortcomings in BCMA design, implementation, and workflow integration encourage workarounds, and that integrating these systems into real clinical work requires attention to how they are actually used in situ, if you want to assure safety.

A workaround is not disobedience; it is a clinical judgment, made at real-time speed, about which failure is safer. A nurse who taps a barcode taped to a doorjamb has decided that the risk of a delayed medication exceeds the risk of a bypassed check, and in an emergency that nurse might be correct. This is not advocacy for the shortcut; heuristics and shortcuts can kill. It is a call to reserve judgment until investigated. Fatigue, overload, distraction, and time pressure explain that far more often than laziness or lack of skill.

The system was built properly. It was tested properly. The work was never watched, so it went around the build, and the purchased safety feature does not work. Without this discipline in the chair, an organization deploys correct technology into a workflow that has not been examined, measures adoption rather than effect, declares the project complete, and never learns that the barcode is on the doorjamb.

Standing and What Happens Without It

Standing is the authority to make decisions and hold the workforce accountable. Informatics fixes can cost real money, take real workforce effort, and cross numerous departments. Retiring an order set or alert means telling an entire service line it is going away. Rebuilding a documentation template means telling revenue cycle and the executive suite their dashboards are about to change. Halting an AI deployment means telling a vendor and an enthusiastic physician that the answer is no, or not yet.

One of the signature failures is the most common of the four, and it is often quiet. The right diagnosis is made for a clinical decision, workflow, or EHR build. It is governed, written down, and agreed upon in the clinical area. A help-desk ticket is entered, the urgency presented. Then nothing happens, repeatedly, for years, and eventually the physicians stop participating because it changes nothing. The Diagnostician has no standing to move the clinical priority into organizational action.

The Rarity of the Whole Prescription

Clinical judgment is common: roughly 1 million physicians in this country have it. Technical fluency in clinical systems is common, but clinical informatics fluency is not. It can be taught; Chapter 10 will have something to say about how badly we have taught it. Workflow engineering is rarer still, the one most often not verified. Workflow is why so many technically sound builds land on a unit and are quietly worked around within a month. Executive standing is a position only the chief executive can give.

All four, in one person, in one chair, is a combination rare enough that most American health systems have never seen it, which is why most do not know what they are missing. It is hard to select for something you have never observed. That combination is the Diagnostician: not the CMIO title, which is free, not the certificate, which Chapter 10 will examine at some length, but the four competencies, held together, in a chair with the standing to use them.

TREATING THE PATIENT WITHOUT A DOCTOR

I have been in these rooms for decades and seen three postures, which each seems defensible from inside the organization, but each is just a way of treating a serious illness without a physician.

Undiagnosed and Untreated

No Diagnostician appointment has been made. Clinical technology decisions are made by the vendor, the technology function, and the quality team, and this never struck the organization as strange, because it has always been that way.

This is more common than the role's impact should suggest: 7 in 10 organizations have the Name (HIMSS, 2019). The other 3 in 10 have nothing at all, and they are disproportionately the community hospitals and rural systems where the margin and workforce are thinnest. The financial exposure per physician unfortunately is the same, the shortages are greatest, and that is where the role would do the most good. There is an honesty in this posture, at least: the organization does not value the Name enough to pay for it, but the gap is visible. Most can point at it.

The Home Remedy

The role exists the way a folk remedy exists. A respected physician has been given a stipend, a day a week, and an invitation to represent clinical interests to IT. The effort is meant well, and the board has been told that clinical informatics leadership is in place.

Chapter 8 described why this is worse than doing nothing: it makes the gap disappear on paper without closing it, and the same setup that keeps the role from working also hides the fact that it isn't. The role is filled by a "rented prescriber" for a defined number of hours a month. They arrive, they are briefed, they offer their view, they invoice, and they leave. The organization has purchased advice and believes it has purchased governance and a plan. Nothing that matters in this book can be done part-time by a Diagnostician with no accountability for the consequences.

Treated as a Minor Complaint

The role exists, but in a diminished capacity. This is the most frustrating of the three, and it is common even in large, sophisticated systems. The person is trained, there is a team, and the whole apparatus is positioned as an operational service function, below the strategic conversation, in the part of the organization that receives decisions rather than makes them.

The tell is the meeting calendar. This person is in the technology governance meeting, not the capital allocation meeting. In the quality committee, not the strategic planning retreat. Told what the organization already voted to do, never asked what should be done. The patient is sick, deteriorating, and being managed by a physician called in after the plan is set.

CLOSING THE COST OBJECTION

Those three previous postures are defended the same way, and the defense is always about money. We cannot justify a full-time executive for this, we cannot carve out that much protected physician time, we cannot move those resources out of information technology, or let us revisit the question next year.

After Chapter 8, that conversation should be over.

The new money is about $1,184,000 for a system with 1,000 physicians. The financial exposure that Function is deployed against is about $840 million a year. To pay for itself, it must recover only $0.14 (new money) out of every $100 of financial exposure.

So, when an organization declines this on cost, a committee of intelligent, well-intentioned executives is declining to spend $1.2 million to address $840 million of documented annual loss, and calling it fiscal discipline.

WE ARE ARGUING OVER A ROUNDING ERROR WHILE THE PATIENT BLEEDS TO death. There is a real objection underneath the false one. Reorganization is genuinely difficult; moving analysts, budget, and reporting lines between executives is uncomfortable, political, and slow. That is a major reason this does not happen enough. But that is a governance and courage problem, not the money problem it has been dressed as. The cost of the Diagnostician stopped being the interesting question. What remains is whether the organization is willing to be courageous and take the step.

CHOOSING THE WRONG DOCTOR

If the Diagnostician qualification is a combination of four things, there are three ways to get the appointment wrong, and American healthcare has been doing all three, at scale, for two decades. All three can be condensed into a failure to exercise *specialty-matching*. As when a brilliant orthopedic surgeon is sent to read a complex abdominal CT scan, healthcare operations are struggling with *specialty-matching*. The available executive roles in modern healthcare have not been trained to treat the illness that the organization has. Each is proficient in their operational specialty. Not one of them has been built for the Diagnostician role this case requires; none of them were presumed to be.

The Wrong Doctor

The Form is assigned to a clinical executive whose training and accountability sit elsewhere.

The chief medical officer (CMO) is a frequent recipient. The CMO is responsible for the practice of medicine in the organization: credentialing, peer review, medical staff governance, clinical policy, the medical executive committee, an enormous portfolio, but not this one. Asking that executive to also govern the decision support build, evaluate AI models across specialties, and design the documentation workflow means asking whether one person can do both jobs at the depth each requires. The exception exists: a CMO trained in clinical informatics is qualified to do the integration. The remaining question is time, and in most organizations time alone makes it unsustainable.

The chief nursing information officer (CNIO) is the second frequent recipient, and this one wants more care, because it is regularly misread as a comment about nursing. It is not. Nursing informatics is a real discipline, doing real work, scoped to nursing EHR practice. The enterprise clinical technology function is scoped to the entire clinical enterprise: nurses, allied health professionals, and the medical staff together. Handing the enterprise resources to a leader trained for the nursing EHR experience is a scope mismatch, and the organization made it, not the nurse.

The data make the next point rather than I do. The KLAS Arch Collaborative finds nursing cohorts stuck at the bottom of the technology maturity curve at more than twice the rate of physician cohorts (KLAS Research, 2026a). Nurses are not the obstacle to this diagnosis, they are its largest casualty. A chief nursing information officer already struggling to obtain basic technologies and maturity for their own domain, then handed an enterprise portfolio, is carrying a mismatch that most other clinicians never have to.

I have delivered this diagnosis to nursing executives who responded with visible relief, because they had not wanted the informatics portfolio they had been assigned and had been saying so for years. I have also delivered it to nursing executives who had held the form and heard the critique as personal. Neither group is the target. The structural mismatch is the target.

No Doctor

This Form is assigned to the chief information officer more often than others, because the historical reporting line was already there: the EHR is technology, the budget sits in information technology, and the decision requires no thought.

The chief information officer (CIO) is responsible for infrastructure, security, vendor management, and the operational stability of every digital service the organization depends on, a large and demanding portfolio that exists substantially outside clinical practice. The same exception applies as with the CMO: a CIO who is also a physician trained in clinical informatics is fully qualified to do the integration, and the same time constraint makes it unsustainable.

Clinical informatics is not a subset of information technology. It is its own discipline, with its own academic tradition, board certification, scholarly literature, and professional society. It is interdependent with information technology but not interchangeable with it. Wherever the responsibility is folded into the technology portfolio, the failure is the loss of contact with, and participation from, the clinical practice the function exists to serve.

Some of this arrangement persists because relinquishing it means relinquishing budget, headcount, and standing, and no executive gives those up eagerly. But this is treating the patient without a license.

The Untrained Doctor

The third failure looks correct from the outside, which makes it the hardest to detect. A physician has been appointed. They have the Name, maybe a budget, maybe even a team. They are competent, well liked, hardworking, and they were never trained for this, because until very recently there was no reliable way to be. Chapter 10 examines how that happened.

The consequence: an untrained appointee cannot read the machine, cannot decipher the workflow, and cannot analyze the data. The clinical decisions drift back to the analysts by default, and the organization believes it has solved the problem.

These three failure modes share something in common. They are problems of *specialty-matching*, not personnel problems like most are treated. The overloaded CMO, the mis-assigned CNIO, and the miscast CIO are called understaffed and ineffective. The unprepared physician is called a bad hire. They all represent a structural failure and guarantee that the next appointment fails in the same way.

SCOPE OF PRACTICE

Medicine solved this problem a century ago and gave it a name: *scope of practice*. We do not litigate whether a given physician specialist is competent for a decision. We define, in advance and in writing, which decisions belong to which credential. That is why the administrator in the opening vignette produced a flinch: every clinician in the room knew the boundary, because the boundary was written down.

Clinical technology has no equivalent, and it needs one. Here are five decisions that belong to a physician and are routinely made elsewhere.

Clinical decision support content and firing logic. What the rule says, when it fires, whom it interrupts, what happens when it is overridden, what the evidence-based recommendation is, where the decision gets made, what frequency and style of alert makes sense: they are all a clinical design choice, and in most organizations are made by people other than the physician.

Order set content and defaults. Order choice, content, and defaults are standing clinical recommendations issued at scale, and should be governed the way any standing order is governed.

Documentation architecture. What a note must contain to be helpful, what the template requires to comply and bill, what elements it pre-fills, what it makes optional. Chapter 5 established that the record is now the evidence for payment, quality, research, and legal defense; what the template asks for determines all four, and all are physician-specific choices.

Deployment and discontinuation of AI models. Whether an algorithm is validated on the right population, monitored across subgroups, and what finding causes it to be switched off is clinical decision authority that belongs here.

Alert burden. Not just the individual alert, but the cumulative interruptive load a physician faces across a shift, which no individual alert owner is accountable for and which goes completely unmeasured.

Five decisions. All are clinical. In most American hospitals, they are being made by roles practicing way above their license. (See Figure 9.1.)

THE DIAGNOSTICIAN'S ROLE · DECISION RIGHTS

Who Writes the Prescription?

Every setting in the building runs on decisions that only look technical.

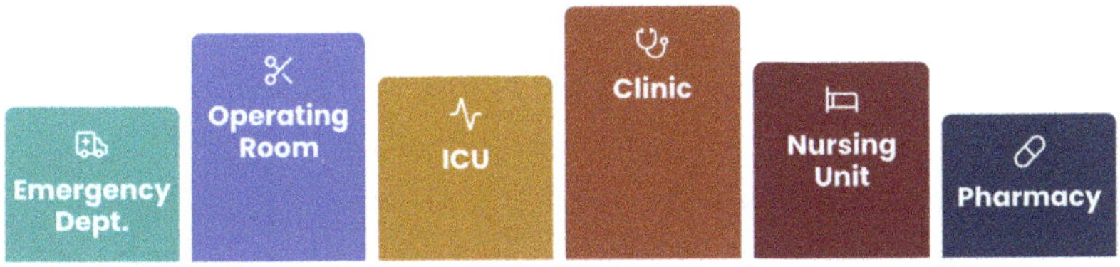

1. **Medication Defaults** — The default dose for a drug that can injure the kidney, and which order sets use it.
2. **Alert Logic** — Which safety alert fires, what it says, and when it should go quiet.
3. **Documentation Design** — Whether the template asks for a finding at all, and how it gets captured.
4. **AI Deployment Safety** — Whether a model is safe to use on this population, in this way, for this patient.
5. **Alert Severity** — Whether an interruption needs a hard stop, or only a soft nudge.

All five are technical settings inside the EHR. All five are clinical decisions. Confusing the two is how patients get hurt.

FIGURE 9.1

Five decisions that look technical and belong to the clinician, in every setting in the building.

BEDSIDE QUESTIONS: THE EXAMINATION OF YOUR DECISION-AUTHORITY BOUNDARIES

1. For your top ten decision support rules, name the clinical leader who approved the current configuration. Not the committee. The person.
2. Who chose the defaults in your three highest-volume order sets, and when were those defaults last reviewed against current evidence?
3. When an algorithm is deployed in your organization, who has the authority to switch it off, and have they ever used it?
4. Is there accountability for the total interruptive burden a clinician experiences in a shift, as a single measured quantity?
5. Write down, in one sentence each, which clinical technology decisions belong to your chief information officer and which belong to your clinical informatics leader. If the two lists overlap or cannot be produced, you do not have a boundary.
6. Walk one unit for an hour and count the workarounds. The taped barcode, the shared login, the parallel spreadsheet, the sticky note on the monitor. Each is a clinical judgment about which failure is safer, and are design defects that could have been caught by watching.
7. In the last year, has your clinical informatics leader been in the room when capital was allocated? Not briefed afterward. In the room.
8. Which of the three postures in this chapter describes your

organization? Not treating at all, treating at home, or treating it as a minor complaint?

9. What is the actual reason you have not built this function? If the answer that comes out is cost, go back to Chapter 8 and check the arithmetic, then find the real answer.

If the answers are concerning, it suggests that the prescriptions in your building are not credentialed, and your organization hasn't flinched.

THE DIAGNOSTICIAN HAS BEEN IDENTIFIED AND THE qualifications named. Clinical judgment, technical fluency, workflow engineering, and standing, in one person, in a chair with real authority. That is the Diagnostician, no matter what Name is on the door.

What remains is why that combination is so rare in a country with over one million physicians and a subspecialty board dedicated to producing it. Chapter 10 examines the profession that was supposed to build this workforce.

CHAPTER 10

A DISCIPLINE IN DISARRAY

THE PROFESSION OF CLINICAL INFORMATICS

EVERY MEDICAL SPECIALTY HAS BEEN ESTABLISHED FOR A REASON. THEY ARE willing to say why they exist and what their discipline should be doing. Their Name, Form, and Function are established. Let us explore clinical informatics as a medical specialty.

A RESEARCH TRIAL IN THE U.S. IS TESTING A NEW ANTIBIOTIC REGIMEN meant to treat appendicitis without surgery. A hospital's medical team enrolls three patients from their floor in the study. One receives no treatment at all, a comparison case. One receives the trial's placebo, a preparation with no active drug in it. One is supposed to receive the full dose, but a calculation error delivers a fraction of what the protocol calls for.

None of the three treatments improved the patient, all three went to surgery. The team logs three cases and three failures, and begins, privately, to doubt the drug.

Every other site in the trial is seeing real results. The difference was never the drug. It was the dose. Two received none and one received little. The difference was undetectable.

THE ROOT CAUSE ANALYSIS

Good hospitals hold morbidity and mortality conferences (M&M) to examine poor outcomes. All hospitals conduct root cause analysis (RCA) after a patient is injured, reconstructing where the care went wrong, how the system failed, or how the patient was hurt despite no detectable error. Aviation and the other high-risk industries follow the same protocol, because an industry that will not examine itself repeats its mistakes.

The case is presented. The team reconstructs the timeline, reviews the details, and names the contributing factors, including the clinician's own errors. Assigning blame or motive is not the focus. The session ends with a plan.

These are some of the most productive hours in medicine.

PRESENTING THE CASE

One clarification before the chart review starts. This book uses *applied clinical informatics* to mean one specific thing, the discipline practiced inside a hospital or health system, at the bedside, in the room where operational decisions get made. Clinical informatics also lives in two other rooms, industry, which employs it to design and consult on clinical applications, and academia, which researches it. I have worked in both. Neither is this book's subject.

Nearly every search for a clinical informatics leader tells the same story.

A health system loses its informatics leader, or decides it needs one. The reasons vary: an EHR go-live, a role to refill, a rebelling physician group, an industry best-practice half-read. The chief medical officer (CMO) and chief information officer (CIO) run the search and produce three candidates. What should this person do? It has not been written down. What should they be able to do? No measurable skill has been defined. Who reports to them? Undecided. Who do they report to? Also undecided. What does clinical informatics mean locally? Several things at once, and nothing, because the organization already employs people at four salary levels under the same name, sitting in IT, in nursing, and in the CMO's department. How will leadership know if this worked? There is no measure, because you cannot measure what has never been defined.

They hire a candidate anyway.

The new executive spends the first several months finding a niche. The organization is proud of its matrix leadership strategy, a phrase mistaken for sophistication. For the new executive it means that there are no direct reports and help will be scarce. There is no informatics operating plan, no budget, no staff, because a calendar full of standing meetings does not require any of them. IT owns the EHR. Revenue cycle owns the budget, the majority of the spend, and the front of the helpdesk queue. Quality is the only group permitted to change clinical content. Nursing dislikes the physician workflow. The physicians have stopped participating, because the person sent to help cannot fix anything.

A year passes. The margin is worse. Regulators visit more often. The clinicians are more burned out. The engagement survey is more brutal. Patient satisfaction is historically bad.

The chief executive begins to wonder whether informatics, and the physician time it consumes, is worth the money.

This is the case. Let us evaluate the facts.

RULING OUT THE EASY DIAGNOSIS

It is easy to blame a slow executive learning curve. But, there is enough blame to go around.

American healthcare executives are hiring informaticists badly or not at all, a fact this book has already shown. The industry itself remains an operational, safety, and financial mess, also already established. The legislation that propelled healthcare into the digital data age created the need for informatics without a plan to supply it. It then mishandled the digital infrastructure, the incentives, the execution, the supporting workforce, and the outcome measurement that followed. Chapter 17 addresses this in full.

The largest contributing factor may sit inside the profession itself. The Name, the Form, and the Function have been an enigma, and the professional board and workforce have stayed silent.

THE STANDARD OF CARE OF A SPECIALTY

A Name, a Purpose, and the Resources to Match

A medical specialty is given a name and a purpose. Every cardiologist completed medical school, an internal medicine residency, and a cardiology fellowship, and that name protects the physicians who treat the heart. A person with an interest in the heart cannot etch "Cardiologist" onto a name tag and assume the role. An employer rarely struggles to hire the Name, support the Form, and use the Function.

The specialty then receives what it needs to do the work. A surgeon gets an operating room and anesthesia support. A hospitalist gets the authority to write orders. A cardiologist gets a catheterization lab. A busy orthopedist gets a physician assistant. No hospital hires a surgeon and then asks them to find their own operating room, or forgets to stock the medicine their physicians will order. It is unthinkable.

An Undisputed Scope

Every specialty also arrives with a defined scope. Orthopedic surgeons do not read abdominal imaging, but they perform knee surgery. Psychiatrists do not perform bypass surgery, but they manage the depressed patient. Radiologists do not round overnight with the hospitalists, but they read the CT scan. What each one does, what the hospital should expect, how they are measured, and what their hiring produces are all known in advance.

The C-Suite Enjoys the Same Stability

The hospital executive role is no less settled. The chief executive officer runs the organization. The chief financial officer owns financial strategy. The chief information officer runs the technology that today supports nearly every other function. Each gets an office, an assistant, a budget, a staff, and a fixed place on the organizational chart. Each has a recognized route into the role, less formal than a residency but real. Each deliverable is understood, the CFO files the audit, the CIO reports system performance, the CNO tracks nursing turnover and survey results. The salaries, the output, and the performance targets for each of these roles are benchmarked and published.

Now set the informatics executive beside these roles. No reserved name. No defined structure. No defined scope. No standard reporting line. No published deliverable. No agreed measure of success. No accepted route in.

That is the finding. The rest of this chapter explains how it came to be true.

WHAT A NAME MEANS

The largest share of blame for what follows belongs inside clinical informatics itself. It belongs to us, because we stayed quiet while three things became true.

The Name. American healthcare has no single working definition for clinical informatics. A large employed workforce carries the title today without the clinical or informatics training the definition requires. EHR vendors use the name. IT departments use it. Nursing departments use it. Individuals put it on a business card. They all believe it is an available label and are not told otherwise.

The Form. There is no standard way to deploy this workforce or the executive who should lead it, not the hiring requirements, the salary, the reporting line, the authority, the staff, the budget, the strategic plan, the scorecard, or the expected return. The role is executed in as many ways as there are organizations doing it.

The Function. There is no common scope of practice, or definition of what the work is. No agreed upon methods, how it gets done. No standard metrics, how you would know it was done well. And, almost no operational literature, meaning whatever it returns has never been published, including what worked or failed.

A profession with no definition, no standard deployment, and no published results is not yet a profession. It is a name in circulation.

This is also why I have spent so much effort insisting the Name itself is not the point. The Form and the Function are what matter, not the label. But, please, give it a name.

THE HISTORY OF CLINICAL INFORMATICS

Before 2013. An educational pathway existed, and a few walked it. In the 1990s, those of us who saw the coming digital age in healthcare trained in medical informatics programs. These programs were built for precisely the competencies this book describes. The National Institutes of Health and the National Library of Medicine funded fellowships and graduate degrees at a small number of universities. The field was not new even then; it dates to 1972, and in its early years supported eleven doctoral candidates across four institutions in the entire country (National Library of Medicine, 2018).

In 2013. The medical subspecialty opened. The American Board of Preventive Medicine and the American Board of Pathology offered the first certification examination in clinical informatics, and physicians could qualify through experience, attested prior education and informatics work, or attested EHR-related experience. That year, 488 sat for the examination and 445 passed (Lehmann et al., 2018a; Detmer and Shortliffe, 2014).

2016. The first fellowship-trained applicants were admitted to the examination.

2017. The board documented a shift. The pass rate fell from 91% in 2013 to 77% in 2017. The examination's difficulty was deliberately held constant, the criteria unchanged, the reviewers unchanged. What changed was who was applying, "clinical domain experts who collaborated with information technology departments to generate artifacts such as order sets, decision support, and documentation" (Lehmann et al., 2018a).

2025. The experience route closed after the 2025 examination cycle, urged by an open letter. One pathway remains, a minimum of 24 months in an Accreditation Council for Graduate Medical Education (ACGME)-accredited clinical informatics fellowship (Turer et al., 2022; ABPM, 2025).

What remains to supply clinical informatics for the U.S. is around 60 accredited clinical informatics fellowship programs, offering around 195 approved training positions, and only around 60% of these positions are actively filled, yielding approximately 50 graduates per year (ACGME, 2025; ABPM, 2025).

WHAT THE CERTIFICATE MEANS

The clinical informatics board examination tests domain knowledge, data structures, standards, decision support logic, and some of the research literature. Those are the right things to test, and a physician who passes it knows the material.

It does not test the practice, and it was never built to. The board treated the experience attestation before the test as the only educational, training, and operational input. The board itself noted that these attestations grew weaker over time.

The examination was not designed to test operational aptitude. It does not measure whether a physician can chair a committee that four other executives would rather run themselves, redesign a workflow that the nurses and physicians live inside, complete an improvement project to a measured result, or hold a clinical position against a chief financial officer while the money is being budgeted.

The fellowship now adds supervised operational work, the right idea, and the only place that work can be assessed formally.

Without blame or bad intention, this section extends the RCA into some of the contributing factors behind America's confusion. Hospital administrations are trying to find clinical informatics's active ingredients. We have made it hard for them to know whether they have been handed a placebo. The certificate states the contents correctly. What it cannot state is whether the medicine will work in this patient, because that was never what it was for.

The skills exist. There are executive CMIOs who have led extraordinary healthcare transformations for years. Building a standard that distinguishes placebo from cure is something the profession owes its patients.

THE DIFFERENTIAL DIAGNOSIS

The people described so far are board-certified clinical informaticists. They are not the only ones using the Name.

Three things first. This will likely irritate some people who hold the title. It describes only how one name became many jobs with almost nothing in common. And, it is an RCA, not an accusation.

Up to this point I have used clinical informatics and applied clinical informatics interchangeably. The distinction between them is only in application, the same discipline practiced in a clinical setting rather than an academic or industrial one.

The groups below have all been called clinical informatics. A title on a name tag, a department that inherited the Name, a short course whose certificate arrived with the name printed on it. Some person went to work under a title already chosen. The failure belongs to a field that defined a term, certified it, and then never defended it.

There are at least 10 ways to end up with this Name. One is the official route, and it is not the one producing most of the titles.

The trained and practicing informaticist. Graduate training built to produce people who would apply it, then 20 or 30 years applying it in operating seats. Some sat the examination, some never did. This is the Function this book describes, known by degree and experience.

The fellowship-trained subspecialist. Two years of supervised operational training and a board examination, about 52 produced a year. The newest functional training.

The practice-pathway certified physician. Certified by examination between 2013 and 2025 on attested experience, which may or may not represent the function, and the certificate does not say. Many of the early degreed informaticists certified this way, and are safe hands. The vast majority in this group have far less experience.

The research scientist. A master's or doctoral degree in biomedical informatics with a clinical focus, training in computing methods, language processing, and study design. It produces excellent research scientists who move the field forward. It is not the operational function, and was never meant to be.

The academic-focused clinical informaticist. A physician holding one of the informatics degrees who teaches, publishes, and trains fellows, sometimes even having the CMIO title with no operational responsibility. The organization gets the title and believes it has the function, and may never learn the difference.

The appointed champion. A respected physician handed the title, a portion of their time, no staff, no budget, and no authority to stop anything. Common in American healthcare, and the placebo from Chapter 8.

The vendor-certified builder. A physician who completed an EHR vendor's certification course, trained in the product rather than the function. It teaches what the software can be made to do, not what it should be made to do.

The clinician relocated into technology. Often a nurse or pharmacist performing IT support under a locally invented title. Essential work, and not the function.

The nurse informaticist. A real discipline with its own training, literature, and certification, not the function, and the largest group of casualties from its absence, trapped in a nursing support role.

The self-declared. No training, no credential, no examination, no supervised practice, just a title from an organizational chart, a business card, or an online profile. Not the function, and not a small group.

Ten routes. One Name.

THE UNLABELED PRESCRIPTION

Clinical informatics has a scientific definition. It has accredited fellowships. It has a board certification issued by the American Board of Preventive Medicine. This puzzled me for years: how can a field with all three let others use the word?

Because none of those three protects a Name. They were never built to.

The board certifies physicians, and that is its entire jurisdiction. It has no authority over what a hospital calls a department, what appears on a job posting, what a vendor calls its training program, or what a person puts on a business card. It certifies people, not language.

Medicine protects its other boarded titles through national, regional, and local standards. The untrained cannot call themselves a surgeon and operate, because the state will stop them and the hospital will not credential them. Clinical informatics is not a licensed practice, it does not involve direct patient care, so nothing in the licensing structure reaches it.

The professional societies never claimed the word either. They defined it in their literature, where names are mentioned, and never defended it in the market, where names get entrenched. The word entered the language unguarded, and the market did what markets do, it attached the Name to whatever needed one.

Compare it to cardiology one last time. One route, fellowship, examination, certification, and a public list of who holds it. A hospital cannot accidentally hire a cardiologist who took a vendor course. No department is called cardiology because it employs an echo technician. The title tells you the training, the training predicts the work, and the market has relied on that for 60 years.

Clinical informatics has a definition, a fellowship, and a board certification, and not one of the three governs who uses the Name. Nothing else in American healthcare does either. The gap has remained open for decades with the assumption that others knew what the role was supposed to do.

The hope is that a book like this, or some regulation, eventually puts a "fence" around it.

Know what the Diagnostician's Form and Function can and should do, and call it any name you like, they are free.

A QUICK HISTORY AND PHYSICAL FOR THE EXECUTIVE READER

With the Name no longer useful, ask four questions instead.

How were they trained?
What have they been accountable for?
What did they actually do all day?
And, were they ever the person allowed to say stop?

Those four answers will place the correct individual on the list in under a minute, and they are the four that predict whether the Function is even present.

Ask them of your own people. The list produces something valuable, and only 2 of its 10 routes are built to produce the Diagnostician. Most organizations will be surprised by what they find they are paying for.

BEDSIDE QUESTIONS: THE EXAMINATION OF YOUR INFORMATICS WORKFORCE

1. Does the person leading clinical informatics in your organization hold a clinical informatics degree or fellowship, with operational experience and executive participation? Can you name the credential without looking it up?
2. Ask that person which of the four competencies in Chapter 9 they were formally trained in, and which they learned by doing. Their answer is more useful than their curriculum vitae.
3. Does your clinical informatics workforce report to that person, with a dedicated budget and a staffing plan?
4. Do you require anything beyond a clinical degree and a willingness to participate before they hold an informatics title?
5. Is there a defined job description and career path for the clinicians doing informatics work in your organization?
6. Are you sponsoring internal candidates into formal informatics training, or relying entirely on external recruitment from a supply channel that has just narrowed by an order of magnitude?
7. Who in your organization can describe, in one sentence, the difference between a vendor-certified analyst and a board-certified clinical informaticist?
8. When you last hired for an informatics title, who wrote the job description, and had they ever done the work?

9. Score your organization for the six Diagnostician and informatics success conditions: education and preparation, funding, positioning, empowering, staffing, and employing. Then ask yourself honestly whether the results you have seen from this role are evidence about the role, or evidence about the dose.

Silence in these answers shows that there is a clinical informatics definition issue in the organization. What is it, what should it do, how much should it cost, what is it worth, and where should it live. An organization cannot resource a role it cannot describe, and it cannot recruit for competencies it has never written down.

THE CASE HAS BEEN PRESENTED, THE TIMELINE RECONSTRUCTED, and the contributing factors to the Diagnostician confusion named. What remains is establishing the diagnosis, which is written down and signed. Part IV commits to the diagnosis.

PART IV

THE DIAGNOSTIC SYNTHESIS, DATA INTO DIAGNOSIS

Every physician knows the moment in a complex case when the information stops accumulating and the synthesis must begin. The history is complete. The examination has been performed. The labs are back. The imaging has been read. The consultants have weighed in. None of these steps, by itself, is the diagnosis. The diagnosis is the disciplined integration of everything gathered into a single, named pathology the treatment plan can address.

Let us apply this synthesis to the modern American healthcare organization. Let us name the disease.

CHAPTER 11
THE WORKING DIAGNOSIS
MAKING A COMMITMENT

A DIAGNOSIS IS A COMMITMENT TO A DECISION, WRITTEN DOWN, WITH THE alternatives ruled out and the confidence stated, so that the disease can be addressed and the treatment team can be held accountable.

A 56-YEAR-OLD MAN ARRIVES IN THE EMERGENCY DEPARTMENT WITH crushing chest pain that started an hour ago. Chest pain like this can come from a dozen sources, and two of the most dangerous look almost identical at the door, a heart attack and a tear in the wall of the aorta. Mistaken for each other, either one can kill the patient.

The physician remembers two rules from training. Common things happen commonly. Deadly things need quick action. A heart attack is far more common, but a torn aorta is the one that cannot be missed, so the physician checks the blood pressure in both arms, asks about pain in the back, and reads the EKG the moment it prints.

The EKG shows a clear pattern of artery blockage. The pressures match in both arms. There is no pain in the back. The physician commits to a heart attack, sends the patient to the catheterization lab, and writes a chart that explains not only what was found, but why the torn aorta that was considered was ruled out, and why.

THE MOMENT OF COMMITMENT

There is a particular moment, after a long workup, when the lead physician is expected to name the diagnosis. The team has presented the facts, the labs have been reviewed, the imaging is on the screen, the differential has been generated, and the colleagues in the room are waiting for one person to write something down.

The job now is to commit, and the act is harder than it sounds. Every experienced physician knows the temptation to defer. Order one more test. Wait for one more consultant. Leave the differential open, so nothing must be ruled out and no choice can be wrong.

That pattern has a name in the diagnostic error literature, and it is its own pathology. D*iagnostic hesitancy* drifts into days of accumulating data that produce no additional clarity, while the underlying condition proceeds unchecked.

The discipline is to commit. Name it. State the supporting evidence. State the alternatives considered and rejected, and why. Calibrate the confidence to what the evidence supports. Write it down.

A committed diagnosis is a hypothesis that can be falsified, and the only thing a treatment plan can be written against. Without it, every subsequent move is empiric, and empiric therapy that is never narrowed becomes the prescribing cascade this book described in the introduction. American healthcare has been cascading for over seventeen years.

The workup is complete. The diagnosis can no longer be deferred.

SUMMARIZING THE PATIENT'S FINDINGS

Stand back and look at the patient as a whole.

The history and physical revealed a multi-system set of signs and symptoms. The clinical encounter, the chief complaint of the entire enterprise, has been degraded by decision support that "cried wolf" until the listener ignored it, by documentation templates never designed for clinical reasoning, and by workflows that consume the cognitive bandwidth of the clinicians the system depends on.

The operational vital signs are abnormal in predictable patterns: emergency department boarding, operating room turnover time, unread inbox message volume, after-hours documentation time, and clinician attrition are all up. The financial markers lag in equally predictable patterns: thin margins getting thinner, climbing denials, quality-linked penalties that arrive as surprises, and regulatory exposure the organization sees only when the auditor reveals it.

The deeper workup confirmed the surface findings and named their causes. The EHR, used as a treatment instead of an imaging modality, was producing artifacts that weren't being corrected, because reading the image was never taught. The chart is mostly recycled words that never get read. Six recurrent problems contaminated the data behind every dashboard and payment calculation, and the organization was auditing none of them. The iatrogenic review found that the technology meant to support clinical care had become, in many configurations, a documented cause of further harm.

The care team examination found that the consults this organization needed were not reliably present. Not the functioning

Diagnostician. Not the funded informatics program. Not the specialty-matching of authority to decision type. None of it. Part III examined the executive-physician who should have been reading all of it, and found only a name on a door, a placebo dose, and a discipline that has never once measured and published what it produces.

The findings are consistent. This is a complex organism with a diagnosable, and diabolical, syndrome.

GENERATING THE DIFFERENTIAL DIAGNOSIS

Before committing, the discipline is to enumerate what else could account for the picture, ranked by probability and severity, with the features that would distinguish them. Six candidates present.

Technology insufficiency holds that the organization has not invested enough in technology, or in the newest tools, and a digital capability gap is producing the symptoms. The outcomes benchmark poorly, so it must be the platform. This is the diagnosis vendors and technology consultants offer most often. Parts I and II do not support it. Investment is substantial in nearly every organization examined, the technology is largely capable, and the gap sits elsewhere.

Clinician underperformance holds that bedside clinicians are not practicing carefully enough, working hard enough, or communicating often enough. This is the diagnosis reached by those whose only available therapies are more education, tighter performance management, and increasing discipline.

Executive incompetence holds that the C-suite is making bad decisions, and strategic missteps are producing the symptoms. This is rarely the case. The executives in most organizations examined are competent inside their own domains.

Environmental attack holds that the regulatory, payer, and demographic environment is adverse enough that the symptoms are inevitable regardless of internal management. There is some truth here. The environment is hostile, and has escalated steadily since

HITECH started, without ever giving the system time to gain its footing. But the measured variation between organizations facing the identical environment is too large for environment to be the primary cause.

Governance fragmentation holds that the organization has the talent, the capital, and the technology, but the synthesis across domains is not happening. Parts I and II strongly support this. The organization's discipline-specific committees function; the cross-disciplinary and cross-executive synthesis does not. Decisions requiring fluency in clinical medicine, technology, workflow, finance, and regulation are made one domain at a time, by leaders fluent in one domain and not the others.

Diagnostic deficiency holds that the organization has never been examined as a whole patient, never had its symptoms read together, never had its workup integrated, and has therefore never received a working diagnosis a coherent treatment plan could be written against. The findings strongly support this as well.

On closer inspection, the last two candidates are not two different diagnoses; they are one diagnosis at two levels of description. The synthesis is not happening because the responsibility to do it has never been assigned. (See Figure II.1.)

THE WORKING DIAGNOSIS · NARROWING THE FIELD

Generating the Differential

Six candidate diagnoses, weighed against the evidence.

Ruled Out · Ruled Out · Ruled Out · Contributing · Primary · Primary

1 Not Enough Technology

HYPOTHESIS Hasn't invested in the right tools.

REALITY Spend is substantial and the technology is capable. The gap is elsewhere.

RULED OUT

2 Clinicians Aren't Trying Hard Enough

HYPOTHESIS Bedside staff aren't working hard enough.

REALITY Clinicians work at or beyond capacity. Most of this traces to bad systems.

RULED OUT

3 Bad Executive Decisions

HYPOTHESIS The C-suite is making bad calls.

REALITY Executives are competent specialists, but only within isolated domains.

RULED OUT

4 Outside Forces

HYPOTHESIS Regulators and payers make failure inevitable.

REALITY Real, but the variation among peer systems proves the dominant factor is internal.

CONTRIBUTING FACTOR

5 Fragmented Governance

HYPOTHESIS Right talent, technology, and capital, but no synthesis.

REALITY Committees function in isolation. Cross-domain calls are made blindly.

PRIMARY DIAGNOSIS

6 Nobody Reads It Together

HYPOTHESIS The symptoms have never been read together as a whole.

REALITY The root of the fragmentation. No one is chartered to synthesize.

PRIMARY DIAGNOSIS

FIGURE 11.1

The differential narrows — three ruled out, one contributing, two describe the same root cause.

RULING OUT THE WRONG DIAGNOSES

A complete diagnostic note documents the excluded diagnoses, both as discipline against premature closure and as preparation for the conversation with the family. In the case of an organization, the family is the board, the medical staff, the executive team, and the community, and they will eventually ask why the diagnosis is this and not something simpler. The good news is that you documented the numbers.

Technology insufficiency is ruled out because additional technology, deployed without governance, has been the dominant therapy for over seventeen years, and the symptoms have progressed throughout. RAND reported that the savings from health IT investment had not materialized, and what was missing was not the technology but the organizational redesign required to convert capability into result (Kellermann and Jones, 2013). Chapter 4 settles it: across 8,657 hospital-years of standardized safety testing, organizations running identical software, at comparable spend, produced measurably different safety performance, and vendor choice explained less than a tenth of the difference (Classen et al., 2020; Holmgren et al., 2021).

Clinician underperformance is ruled out because clinicians in nearly every organization examined are working at or beyond capacity on every objective measure. The literature since the 1999 Institute of Medicine report has been unwavering: bad systems, environments, and processes explain most of these symptoms (Kohn et al., 1999; IOM, 2001). Individual incompetence exists and should be confronted seriously, but it belongs in the outlier column where it has always belonged. Nearly 2 hours on the EHR for every 1 hour of direct patient contact (Sinsky et al., 2016). A usability score of 45.9 out of 100, graded F, with every point of improvement associated with lower odds of burnout (Melnick et al., 2020). Override rates on drug safety alerts running between 49% and 96% (Van der Sijs et al., 2006).

A nursing workforce inventing 15 distinct workarounds, from 31 distinct causes, to keep a barcode system safe (Koppel et al., 2008). Not the findings of a workforce underperforming. The findings of a workforce compensating.

Executive incompetence is ruled out on the same evidence. The failure is not inside any single domain. It is in the space between them, and that space has a shape. Health information technology operates across eight interdependent dimensions, from hardware and clinical content, through the human-computer interface, people, workflow, internal policy, external regulation, and measurement, and a failure in any propagates through to the others (Sittig and Singh, 2010). Every dimension has an owner on the organizational chart; the interdependence has none. A decision requiring fluency in several domains at once gets made inside whichever meeting it happened to land in, competently from one point of view. Not incompetence. It is the lack of Diagnostician synthesis discussed in Chapter 7 and the specialty-matching failure described in Chapter 9, where every consultant writes excellent notes that are never assembled.

Environmental attack is a contributing factor, but needs some clarification. There is no doubt that the regulatory, payer, and demographic environment is adverse. It is also broadly adverse, affecting all clinicians and sites taking care of patients. Across 8,657 hospital-years of standardized safety testing, advanced decision support scores varied immensely between locations using the same test, the same commercial software, and in many cases the same vendor (Classen et al., 2020). And, inside a single building: one platform, one implementation, one contract, one regulatory environment, and 12% of nurses report an elite experience of the record against 22% of physicians (KLAS Research, 2026a). So, while the regulatory and payer environment affect the industry similarly, the results vary widely. Something outside of the environment appears to be the cause.

Governance fragmentation and diagnostic deficiency are the same finding from different resolutions. Fragmentation is what the absence of synthesis looks like from the organizational chart.

Deficiency is what it looks like from the bedside. One vacancy described twice.

A physician commits when the remaining candidate explains every finding, and responds to treatment aimed at it. This one explains the findings. Capable technology producing incapable results. Competent executives producing incoherent decisions. Exhausted clinicians compensating for a system never designed for them. Contaminated data that is never audited. Alerts that never get pruned. Money that can't be traced backward. That is what a domain looks like when it is governed in fragments and never synthesized into a diagnosis.

This one also responds to therapy. Chapter 7 documented what happens when the synthesis occurs. At gynecologic oncology tumor boards, the treatment plan changed, the diagnosis itself was shaped, and the decisions were carried out (Lee et al., 2017). In breast cancer, multidisciplinary work improved survival, and reduced the variation in survival between hospitals (Kesson et al., 2012). No new drug arrived. No new operation was invented. A group of clinicians began meeting to produce one plan instead of several, and women lived longer.

Variation between organizations facing the same environment is what ruled the environment out. Reducing that variation is precisely what the cross-disciplinary and cross-executive synthesis function has been documented to do. The diagnosis explains the case and responds to the treatment.

It is time to write it down.

THE WORKING DIAGNOSIS

Diagnostic deficiency, in the form of an absent or under-constituted Diagnostician, manifesting as fragmented governance across clinical, technical, operational, financial, and regulatory domains, with downstream consequences across patient safety, clinical quality, reimbursement, regulatory posture, workforce retention, and unwarranted clinical variation.

Four features of this diagnosis deserve to be named.

It is primary rather than secondary. The other symptoms, the declining margin, the regulatory penalty, the safety event, the attrition, sit downstream of the diagnosis rather than beside it. Treating the downstream conditions without treating the upstream cause is the prescribing cascade.

It is progressive without treatment. Cognitive load grows. Data drift accumulates. The artificial intelligence footprint expands without governance. The financial markers deteriorate. Variation widens. The workforce thins. Every year without treatment makes the next year's treatment more expensive and the recovery longer.

It responds to a specific therapy, and the dose is now definable. Not a platform, not new information, not another initiative. A trained Diagnostician, resourced and empowered, running a defined method on a continuous cadence. Chapter 8 specified the dose and priced the therapy. Chapter 9 specified the competencies the prescriber must hold. Chapter 10 explained why so few people hold them.

And, it is survivable. Nothing in this diagnosis is terminal, and a meaningful share of it does not merely shrink under treatment, it heals. A disease with a named cause, a priced therapy, and a curable element is not one that should be left untreated.

TREATING THE PATIENT UNDER UNCERTAINTY

A diagnosis becomes a decision here, and this chapter must be more careful than any other place in the book.

I cannot show you a controlled trial. Chapter 8 explained why: not that the therapy ever failed, but that it was never named, and you cannot run the trial of a treatment that was never ordered.

We are in the position every physician knows how to handle. A patient with a condition you cannot confirm, needing a test that does not exist. A treatment comparatively cheap, low-risk, mechanistically sound, aimed at a burden that is enormous and documented. No certainty available before you must decide.

You do what the training taught you: establish the threshold, ask how small the benefit could be and still justify treating, then ask whether the truth could plausibly be smaller than that.

Chapter 8 established the threshold. The Function must recover \$0.34 (full program cost) out of every \$100 of financial exposure to pay for itself entirely, and \$0.14 if you count only the money that is not already being spent. A conservative reading of the published literature puts the recovery at \$2.00 per \$100.

To rule this out, you must believe the true recovery is below \$0.34 in \$100 of financial exposure. That is against an exposure of \$840,000 per physician per year, in a field where the gap between average and top-decile performance is large.

You are entitled to hold that position. What you are not entitled to do is hold it quietly. In seventeen years no study in this field has published a recovery rate, high or low. There is no number in the literature to defend the current course, only the absence of a measurement people assumed had been taken elsewhere.

That leaves two honest positions. Treat, and measure what you get: constitute the role to address the eight categories of loss Chapter 8 named, measure the outcomes it priced, publish the result whichever way it falls. Or rule it out formally: run the study Chapter 17 specifies, prove that the intervention and method are wrong, and ask the Diagnostician to stop asking for the executive chair.

There is a third position, the one American healthcare occupies today. Deploy the Name in numerous organizations. Resource it appropriately in almost none. Measure nothing, and treat the absence of evidence as evidence of absence.

This is healthcare, as a patient, maintained on a sub-therapeutic dose, never titrated, never leveled, whose chart records that the drug was tried and did not work.

WHAT THE WORKING DIAGNOSIS PREDICTS

A correct diagnosis predicts the future: what the patient looks like a year from now if treatment starts, and what the patient looks like a year from now if it does not.

Untreated, the course is continued progression, uniform and documented. The board will, or should, eventually ask harder questions. Currently the questions only produce more committees, the committees produce more activity, and the activity does not produce recovery.

Treated, the course improves on a timeline that requires consideration. The temptation to promise faster results is enormous, and it is how this argument can be lost in the first year.

That timeline does not make this diagnosis popular. That doesn't change the fact that it is the only pathway leading to healing.

The Full Dose Has Never Been Given

The executive priority, the full backing, the right Diagnostician, the full staff, the full empowerment, at the same time and sustained. The only national count approximating this book's topic asked whether a health information leader is even employed. Those organizations said yes to a question about employment rather than authority (HIMSS, 2019).

So, we start with an industry where nearly 1 in 3 didn't even try to hire the Name. Seven years since this study result, no attempt has been made to measure whether the 7 in 10 roles have been impactful, which suggests disinterest: health systems tend to publish what works.

In two decades, this literature holds only one serious institutional examination, from one academic medical center (Kaushal et al., 2006). The market says the same thing. This country fills only 60% of clinical informatics fellowship positions, a shortfall addressed further in Chapter 16 (ACGME, 2025). Organizations competing hard for a scarce capability do not leave 40% of the supply unsought.

Every measure you explore reflects an absence of the question. That is what a Diagnostician finds when the event itself never happened.

THE CHARTED NOTE

Every diagnosis that matters gets written down in a form that can acted upon.

Patient: the American healthcare organization.

Presenting complaint: degraded clinical encounters, deranged operational vital signs, deteriorating financial markers, and a workforce leaving faster than it can be replaced.

Working diagnosis: diagnostic deficiency and fragmented governance, secondary to an absent or under-constituted Diagnostician. The adverse environment is a contributing factor.

Confidence: high, on the strength of convergent evidence, the exclusion of three alternatives, and proof of response to the treatment.

Financial exposure: approximately $840,000 per physician per year across eight categories that clinical informatics can reach.

Therapy: a trained, funded, positioned, empowered, staffed, full-time Diagnostician, with the four competencies, running the method on a continuous cadence.

Cost of therapy: approximately $1,184 per physician per year in new money.

Threshold for benefit: recovery of $0.14 (new money) per $100 of financial exposure.

BEDSIDE QUESTIONS: THE SYNTHESIS OF YOUR ORGANIZATIONAL DIAGNOSIS

1. Do you recognize your own organization in the six candidates in this chapter? Which one would your executive team have named before reading it, and which would they name now?
2. Does your organization have a working cross-disciplinary and cross-executive synthesis function, or does it have five excellent specialists and no lead physician?
3. Has your organization ever committed a diagnosis of this kind to paper, with the alternatives ruled out and the reasoning documented?
4. Run the threshold question on your own numbers. What would your informatics Function have to recover, as a fraction of your own financial exposure, to pay for itself? If that figure cannot be produced, you have not evaluated the decision.
5. Which of the three positions does your organization occupy? Treating and measuring, formally ruling it out, or maintaining a subtherapeutic dose and recording that the drug was tried?
6. Are you early to this, or are you waiting for the technology investment to resolve it on its own?

If you cannot answer these to your satisfaction, your organization has no obvious diagnosis to treat. The first step is not a purchase. It is an examination of the patient.

THE DIAGNOSIS IS WRITTEN DOWN. THE DIFFERENTIAL IS closed. The alternatives have been ruled out with evidence rather than instinct, and documented.

THE THERAPY IS NOT COMPLICATED. It is not a platform, not a new idea, not another initiative. One properly trained Diagnostician, resourced and empowered, running a defined method on a continuous cadence. Every prediction in this chapter turns on whether leadership has given them the authority to act. Let us evaluate some of the modern treatments available.

PART V

THE THERAPEUTIC ARMORY

The diagnosis has been committed in writing. The treatment plan begins. We now characterize some of the major therapeutic modalities available to the modern Diagnostician: clinical decision support (CDS), AI and ambient intelligence, telehealth, and the broader digital interventional toolkit. We use the rigor a physician brings to a new class of medication: indication, dosing, expected effect, monitoring, adverse effects, drug interactions, and discontinuation criteria. When aimed at our diagnosis, each becomes a powerful instrument in modern healthcare.

CHAPTER 12
CLINICAL DECISION SUPPORT
THE IMPACT OF CRYING WOLF

Think of clinical decision support as the automatic reflexes of a modern hospital, translating incoming patient data into clinical recommendations. Like a knee that jerks when a doctor strikes it, these systems fire warnings when something in a patient's chart is amiss. A reflex test teaches a doctor something only when performed correctly and read against a known healthy twitch. Decision support can be done correctly, incorrectly, or not at all. The logic must distinguish the abnormal and dangerous from the normal and trivial. Calibrating that reflex is the work of informatics.

A PHYSICIAN NEARING THE END OF A 24-HOUR CALL SHIFT IS ADMITTING one last patient before they can finally go home. The chart opens, and so does the first alert. The orders are entered and the rest of the alerts flood in. A dose slightly outside the usual range for this drug, though correct for this patient's weight and kidney function. They click past it. A second alert follows, a drug interaction between two medications that, in practice, rarely causes a problem. Clicked past as well. Two more alerts ask them to remember to complete a fall risk form and a discharge planning note, neither one about the patient in front of them right now.

More alerts follow, an allergy check already documented, an unrequested formulary substitution, a wellness screening reminder that has nothing to do with tonight. They have seen versions of all of them a hundred times this week and click through each one in the time it takes to read the words.

One of the alerts in that stream is a duplicate therapy warning, the same kind that has fired and meant nothing for the last twenty patients. This time it is flagging a second blood thinner, already ordered by another team earlier in the admission. It looks exactly like the others. It is dismissed exactly like the others. They go home. That alert's meaning won't be obvious until the bleeding starts.

THE ALERTS THAT CRIED WOLF

There is a story informaticists tell about the moment they realized their decision support was broken, and it gets retold often because it keeps happening the same way.

An informaticist is watching a physician work in an emergency department or an intensive care unit, and a stream of alerts, warnings, and advice crosses the screen. They include drug interactions, allergy checks, age-based dose warnings, duplicate therapy notifications, formulary alternatives, evidence-based suggestions, and regulatory reminders.

The physician dismisses them with a click, all of them, without reading any of them. The click has become a reflex, the way a person brushes a fly off a sleeve. The physician does not pause, does not register that the alert was there, and moves on. The informaticist watching already knows the literature.

Override rates for drug safety alerts run as high as 96% depending on the alert and the setting (Van der Sijs et al., 2006). Numerous drug interaction alerts in commercial systems bear low or negligible clinical significance (Phansalkar et al., 2013). For every alert that matters, 20 or more meaningless ones fire alongside it. The reflex to dismiss the entire lot is a reasonable adaptation to a system whose support has cried wolf until the listener stops listening.

The alerts that are firing usually are not even aimed at the orders most likely to seriously harm or kill, a finding confirmed at the national level and returned to later in this chapter.

The informaticist is watching for the rare alert in that dismissed stream that mattered. The drug interaction that will harm a patient this week. The dose warning that would have stopped an adult sedative dose in an elderly patient. The progressing sepsis the chart's own data already show. The care gap that, closed, would have been worth real money in quality-linked payment.

The wolf is real. The physician stopped listening for him long ago, because the system cried wolf too many times. And, that failure belongs to the designers alone.

This chapter is about how a Diagnostician repairs that reflex. It is about how the diagnostic instinct of an organization, properly calibrated, becomes a reliable instrument in the treatment arsenal.

THE INDICATION FOR CLINICAL DECISION SUPPORT

Decision support is indicated wherever a clinical decision is being made repeatedly, at volume, in real time, under time pressure, with an evidence-based right answer that a busy clinician does not have in front of them.

That is the whole indication, and not what most organizations use it for.

The five rights framework, most often credited to Osheroff and colleagues, holds that effective decision support delivers the right information, to the right person, in the right format, through the right channel, at the right point in the workflow (Osheroff et al., 2024). All five must be present. A failure of any one degrades the effect. A failure of two or more produces the override reflex above.

Kawamoto and colleagues, in a systematic review now over 20 years old, identified four features that predict whether decision support changes practice: it must arrive automatically, as part of the workflow rather than a separate action; it must deliver a recommendation rather than an assessment; it must arrive at the time and place of the decision; and the computer must generate and push it rather than waiting for the clinician to go looking for it. Support satisfying all four improved clinical practice in 94% of trials (Kawamoto et al., 2005).

Writing the Prescription

You do not have to be a Diagnostician, or work in clinical informatics, to notice that almost none of this work is being done in most organizations, not before go-live, not before adopting new CDS, and not on a targeted cycle. Ask for the date an alert was installed and the date it was last reviewed. The answers will surprise you.

Throwing a meaningful alert draws on every informatics skill named so far: understanding the technology, the clinical situations and roles, the clinical workflow, and holding the authority to call subject matter experts to a meeting and change how the system functions.

Right information means high-quality data entry, in a form the computer can evaluate and check against evidence-based logic, and an algorithm built to deliver a reliable answer.

Right person means knowing who should receive the alert and why: the clinician weighing the decision, the one ordering, the one administering, the one holding the relevant license, rather than the default of sending it to every clinician who opens the chart, hoping that somebody, somewhere responds.

Right format means matching the urgency of the message and the desired response to a design the targeted user will read, whether that is a hard stop, an immediate flash, or a banner passively encountered elsewhere in the chart. It also means pairing that design with a response that fits the warning: a button, an order, a structured document, a calculator, a dashboard element.

Right channel means understanding whether the clinician will be logged into the chart when the alert fires, and if not, whether a call, a text, or a nurse alert reaches them in time; the assumption that they will simply be there is a frequent cause of failure.

Right point in the workflow means knowing the workflow the computer forces against the one the clinician actually executes, which takes observation with a stopwatch. A great deal of clinical charting happens after the action was taken, sometimes even after the patient has gone home. An alert thrown at that point is a historical fact and prevents nothing, and a sign on the wall often beats a retrospective warning in the chart.

Arriving automatically means understanding, in full, what data must be available and in what format, what happens when it is not, what clinical action triggers the analysis, what evidence underlies it, how the result is computed, and which clinician, workflow, and device it targets.

A recommendation, as distinct from an assessment, requires an evidence base, expert review, and local clinical vetting before it reaches the care environment. The fastest way to produce non-compliance is forcing a new behavior on an expert clinician without notice. It also requires tools tied to the recommended action itself, whether that is discontinuing a medication, changing a dose, giving more fluid, or documenting a new problem, rather than the older habit of surfacing information and asking the clinician to stop, switch screens, and resolve it unaided.

And, pushing, rather than seeking, means handing the clinician everything needed to understand and comply at the moment the alert fires; the instant they must open a dictionary or go confirm the literature, their attention is gone. If a message read hours later would serve just as well, consider not turning the alert on. (See Figure 12.1.)

THE THERAPEUTIC RESPONSE · CLINICAL DECISION SUPPORT

The Five Rights of Decision Support

Miss any one, and the alert becomes noise the clinician learns to ignore.

	Right	Description
1	**Right Information**	Evidence-based logic, built on data clean enough for a computer to trust.
2	**Right Person**	Sent to the clinician making the decision, not everyone who opens the chart.
3	**Right Format**	A recommendation, not just a warning, matched to how urgent it is.
4	**Right Channel**	Delivered where the clinician actually is, not where the system assumes.
5	**Right Workflow Point**	Arrives before the decision is made, not after the care is already given.

All five must be present. Miss one, and the effect degrades. Miss two, and clinicians stop listening.

FIGURE 12.1

The five conditions that decide whether decision support is heard or ignored.

EFFICACY

Kwan and colleagues pooled 108 studies reporting 122 trials, covering 1,203,053 patients and 10,790 providers, the largest and most careful accounting of what decision support delivers.

Across all of it, decision support increased the proportion of patients receiving recommended care by 5.8% (Kwan et al., 2020). That is the average effect of the most widely deployed clinical technology intervention in the world. In the 30 trials reporting actual clinical endpoints rather than process measures, the improvement in patients reaching guideline targets was a median of 0.3%. Any person selling decision support on the promise of transformation should be shown those two numbers and asked to respond.

The top quartile of studies produced improvements ranging from 10% to 62%, the same intervention class, the same literature. Some programs barely moved practice; others moved it by 62%. The authors looked for what explains that spread and could not identify what separates the top of the range from the bottom.

The variables are informatics and governance. The studies at the bottom were measuring uncurated alert libraries, firing outside the workflow, delivering assessments instead of recommendations, built against guidelines that had moved on. The studies at the top were measuring the same technology under a discipline the paper had no instrument to score, because the instrument had never been built. The predictors are undefined because the field has been measuring the drug and not the dose.

AIMING AT THE WRONG TARGET

A second finding belongs beside Kwan's, and it comes from a national evaluation rather than a trial.

Co and colleagues used the Leapfrog computerized order entry evaluation tool on hospitals tested in both 2017 and 2018. They separated the results by consequence: performance against fatal orders, meaning orders known to have killed other patients, and performance against nuisance orders, meaning orders that trigger an alert without meaningful clinical risk.

Between those two years, overall performance rose. Performance against fatal orders improved slightly. Performance against nuisance orders did not change significantly at all. The authors' conclusion is the sentence I would put in front of every board in the country: hospitals are not targeting the deadliest orders first, and some of the best nuisance-order scores come at the cost of alert fatigue (Co et al., 2020).

The problem, then, is not simply that there are too many alerts. It is that the alerting effort is not aimed at anything in particular. American hospitals are catching the trivial issues about as reliably as before, and catching the lethal ones only slightly better. That means the enormous interruptive burden physicians face has not even been pointed where it would save the most lives. That is not a technology failure. Almost all of those hospitals own a system capable of firing on fatal orders. The failure is prioritization, which is exactly what a governance committee is for.

Put Kwan and Co together and the picture resolves. The effect size at the top of the published range runs more than 10 times the average, and that gap has never been explained. The alerting effort is not aimed at the highest-consequence decisions. Both findings describe the same missing thing, and it is not software capability.

THERAPEUTIC FAILURES

Most organizations that ask me to look at their decision support present with the same complaints: override rates are high, clinicians complain, quality metrics are not moving, and safety events have informatics root causes.

The differential narrows, in nearly every case, to one or more of five conditions.

Misgovernance syndrome

The first and most common. There is no standing informatics committee with authority over content. The library has accumulated alerts the way a drawer accumulates orphan socks, additively, without curation, with no owner, no review schedule, and no removal pathway. The result is hundreds of low-value alerts firing thousands of times a day, drowning the few meant to matter.

The treatment is not more alerts. It is systematic review and rationalization of the existing library, governed by a physician-led committee and informatics with explicit authority to deactivate, modify, and rebuild. The committee's first act is triage: sort the library by consequence, the way Co and colleagues sorted the national data, and find out how much of the interruptive budget is currently spent on orders that cause no harm. In most libraries, the answer is most of it.

The work is unglamorous. The effort is immense. It is also the single highest-yield intervention available in this chapter.

Workflow-design support

The alert is technically correct and clinically appropriate, but it fires at the wrong moment: after the order is signed, before the relevant data exist, on a screen the clinician is not reading. Kawamoto's workflow criterion is missed, and the alert is effectively invisible.

The treatment is workflow co-design. Informatics and relevant clinicians rebuild alerts to fire at the moment the decision is made, on the part of the screen the clinician is attending to. This is not a vendor-default activity. It requires the Diagnostician to watch the work, the third competency of Chapter 9. Most organizations assume it is happening somewhere else rather than staffing to assure it.

Recommendation-free alerts

The alert reports that a problem exists without saying what to do about it. The patient may be at increased risk, consider alternative therapy, documentation appears incomplete; all of those are simple assessments. They consume attention and deliver nothing actionable, the second Kawamoto criterion failed.

The treatment is to rebuild every alert around a defined recommendation. Recommend a creatinine within 72 hours. Recommend metoprolol succinate 25 mg daily. Recommend a pain assessment in the next 30 minutes. When the alert contains the recommendation, the clinician's only task is agreement or disagreement, a far faster cognitive operation than generating a response from scratch. Acceptance rises. Overrides fall.

Static-library support

The library of active alerts may have been built three to five years ago, against guidelines that have since changed, drug data since revised, and organizational priorities that have since moved. They haven't been touched since.

The clinician is being asked to weigh an evidence base that was current several years ago. The more sophisticated clinicians notice the recommendation is wrong and stop trusting the support entirely, including the parts that are correct. The less experienced have not noticed, and data inside their own system are misleading them.

The treatment is continuous content governance rather than a one-time review. A standing committee that meets monthly, a library mapped to the guidelines it depends on and updated when those guidelines update, interaction logic calibrated against current pharmacology and the actual formulary rather than a vendor list from a previous decade. The shift from IT project to clinical capability is the difference between a library that ages gracefully and one that ages into irrelevance.

Shadow decision support

The fifth, often unrecognized and growing fast because of AI. The organization has deployed several decision support engines: the EHR's native rules, a third-party documentation improvement service, a population health engine, and now an AI model or four. They operate without coordination. The clinician receives alerts from all of them, frequently duplicative, occasionally contradictory, with no evaluation, no editorial discipline, and no accountable owner of the confusion.

The treatment is to declare an editorial authority. One clinical informatics and governance function that owns the whole process, decides which engine fires in which context, and maintains coherence across everything the clinician sees. Without that discipline, every additional engine makes the override problem worse rather than better. Chapter 13 will show that the newest engines make it worse faster.

CALIBRATING THE DOSE

The decision support work follows the same discipline as any other clinical problem. The history: what is the library currently doing? The examination: which alerts fire, how often, with what override rate, in which contexts? The differential: which of the five conditions above is operative, and is it more than one? The workup: what evidence distinguishes them? The plan: what gets deactivated, redesigned, or built? The follow-up: re-measure the override rates, the near misses, the clinical outcomes, and what the clinicians say.

A mature program runs that cycle continuously. The committee meets with regularity. Override dashboards are reviewed every time. Safety events and near misses are mapped back to the specific alert and the coverage that should have caught them. New evidence triggers a content update inside a defined window. Clinician feedback is solicited, ingested, and acted on. The library is treated as a clinical content asset, owned and tended like any other.

ADVERSE EFFECTS

Every agent in this library has them.

Alert fatigue is the primary one, dose-dependent the way drug toxicity is dose-dependent. More alerts produce more overrides, more overrides produce a reflex, and the reflex generalizes to the alerts that matter. An organization that answers a safety event by adding an alert has, in most cases, made the next one more likely.

Automation bias runs the opposite direction: a clinician who trusts the computer support will follow it when it is wrong. The same interruptive system produces both fatigue and uncritical acceptance.

Workflow displacement is another: an alert placed to satisfy an administrative requirement rather than a clinical need moves work rather than removing it, usually onto nursing.

And, time displacement: every interruptive alert costs a fraction of a physician's finite daily attention. Chapter 8 priced the physician day at roughly $180 billion a year nationally. Decision support is one of the few interventions that can add to that figure as easily as subtract from it.

DISCONTINUATION CRITERIA

This is the neglected discipline in most organizations, the one that separates a governed library from an accumulating one.

Rank the entire library by consequence before retiring anything. Which alerts stand between a physician and an order that could harm or kill the patient, and which stand between a physician and an inconvenience? Co and colleagues found nationally that those two categories are treated almost identically. That means the ranking has never been done in most organizations. Do it once and the retirement list writes itself.

Retire any alert whose override rate exceeds 90% and cannot be redesigned below it. Dismissed 9 times out of 10, this alert is not providing safety; it is just recording that the alert fired and the clinician ignored it. Retire any alert whose underlying guideline has been superseded and not replaced within the review window. Retire any alert that fires outside the workflow and cannot be moved into it. Retire any alert that reports without recommending. And, retire any alert not defended by name: put it on a screen in a governance meeting and ask who owns it. Silence means it has no clinical sponsor.

A committee with the authority to deactivate is the entire mechanism. A decision support program run by a Diagnostician who knows how to address the accumulation burden.

WHAT DECISION SUPPORT CAN RETURN

Decision support reaches 4 of its 8 categories, worth roughly $660 billion of the $867.5 billion total, the widest reach of any agent in this formulary.

Harm, and what follows it: $112 billion of financial exposure, the largest and most direct effect documented. Order entry with well-governed support has cut serious medication errors by half or better in the published trials. Kaushal's team attributed the largest single component of a $28.5 million 10-year return specifically to the level and type of decision support available (Kaushal et al., 2006). Medication is only one aspect of harm risk, and harm is a target we can mitigate substantially but not eliminate.

Care we did not need to give: $103 billion of financial exposure. The defaults in an order set: medications, equipment, and therapy are usually what gets ordered, and those defaults are being chosen poorly today. Every order needs to be addressed seriously, with appropriate-use criteria delivered at the moment of ordering. Many of these criteria are published in a policy that no clinician will ever read, even if they could find them. This is the mechanism and it can be mitigated substantially.

Paperwork that wastes time: $265.6 billion of financial exposure, the largest category in the ledger. A quality measure handled at the point of care rather than abstracted six weeks later is partly a cure rather than a mitigation. The subset of measures computable from discrete fields can drive manual abstraction to zero, and Chapter 8 counted that in the curable quarter.

And, the half-day that doesn't count: $180 billion of financial exposure, the line where decision support can move in either direction, and most organizations are currently moving it the wrong way. A rationalized library returns clinician attention; an accumulating one consumes it. At override rates running as much as 96%, most of what a typical library produces today is interruption without effect (Van der Sijs et al., 2006). The price of admission is the smallest in the formulary. The library already exists. The engine is already licensed. The alerts are already firing. What is missing is a chartered committee, and a Diagnostician with the authority to turn things off.

That is the return when you convert the most-ignored feature in your record into your most reliable first-line agent.

BEDSIDE QUESTIONS: PRESCRIBING AND DOSING YOUR CDS

1. What is your alert override rate, by alert and in aggregate? Who reports it, and who is accountable for it?
2. Sort your alert library by consequence. What fraction of your total alert volume fires on orders that could seriously harm a patient, and what fraction fires on orders that could not? If the question has never been asked, assume your answer looks like the national one.
3. Does your organization have a chartered committee with explicit authority to deactivate an alert? Name the last alert it retired.
4. When was your library last reviewed in its entirety against current evidence?
5. What proportion of your alerts deliver a recommendation

rather than an assessment? If this has never been counted, that proportion is lower than you think.

6. Do your alerts fire at the moment of decision, or downstream of it?
7. Have you mapped your recent safety events back to the decision support that should have prevented them?
8. Who owns the integrated experience across your record's native rules, your third-party engines, and any model-driven support? If the owner doesn't exist, your shadow support problem is already worse than you think.
9. For every alert firing more than a thousand times a month, is there a physician who sponsors it?

Unanswered, they describe poorly calibrated decision support. That puts you in the company of most organizations. You need a Diagnostician with the training to know which alerts should go and the authority to remove them.

DECISION SUPPORT, CALIBRATED AND GOVERNED, IS ONE OF the most reliable first-line agents in the formulary. It is a tool you already own in the EHR. It can change clinical behavior at scale without adding capital or staff, and the physician-leader who governs it well has done something that sounds modest and is not. They have made the right thing the easy thing to do. Let us now evaluate AI as a therapy.

CHAPTER 13

ARTIFICIAL INTELLIGENCE IS HERE

READING THE LABEL

ARTIFICIAL INTELLIGENCE IN CLINICAL CARE HAS REACHED THE POINT IN ITS life cycle where it is no longer experimental and not yet routine. Like any new class of therapy in that phase, it needs a physician who can name the indication, set the dose, anticipate the adverse effects, and stop it if the harm exceeds the benefit.

A SALES REPRESENTATIVE VISITS AN ORTHOPEDIC SURGEON TO introduce a new hip replacement implant. The pitch is polished: easier to place during surgery, expected to last longer, and associated with less pain afterward. The surgeon takes the pamphlet home to read before the next case.

Buried in the fine print is the reason the company can make those promises so quickly. The implant never went through a clinical trial, because it looks enough like its predecessor to skip that step. The better performance is credited to a new surface coating, which the company will not describe, because it is proprietary. The composition, mechanism, durability, and toxicity are unknown outside the company, along with whether it even works.

The surgeon's next case for this procedure is a patient who is younger and more active than most; few patients this age have ever received this implant before. Around town, colleagues are already switching over, eager for the newest hardware on the market. The surgeon reads the pamphlet once more, unconvinced. If a complication shows up years from now, this patient will be one of the first people to find out whether the coating works, in a patient like them, from a surgery they are paying for.

THE DRUG THAT ARRIVED BEFORE THE TRIAL

I was in a room recently, when a health system chief executive watched an ambient documentation demonstration and said, out loud, that this would solve the burnout problem.

That executive was not wrong that the technology was promising. They were wrong in the way a patient is wrong when they tell you a diet pill cured their diabetes. Weight loss is the cause of the tighter control; the diet pill was one intervention along with exercise. Removing unproductive busywork reduces burnout; ambient documentation is one intervention among many.

There are moments in the history of medicine when a new class of therapy arrives and changes the standard of care so completely that the practice before it looks unrecognizable. Antibiotics. Insulin. Monoclonal antibodies. They all arrived imperfectly. Each was misused before it was used well. Several produced harm before they produced healing, sometimes the harm was catastrophic, more often it was quiet.

Artificial intelligence in clinical care is at one of these moments. I am not passing criticism but drawing careful attention. I am describing a class of agent that is already in the building, already in the workflow, and already in front of patients, and has never been dosed.

Physicians who practiced medicine through one of the great therapeutic transitions recognize the shape of this. It is exciting, dangerous, and demands rigor. And, more than any other modality in this book, it requires a Diagnostician who can think like a physician about this new "drug".

What exactly is in the package?

READING THE LABEL

As a fan of the new possibilities, I start with what the regulatory agency itself has said, because most executives believe it says more than it does.

The Food and Drug Administration (FDA) maintains a public list of AI-enabled medical devices granted marketing authorization. Through the end of December 2025, the list carried 1,451 devices, going back to 1995. Radiology accounts for 1,104 of them, roughly 3 in 4. Hundreds more have been authorized in 2026 and the pace is accelerating (FDA, 2026).

That large number is usually read as reassurance, evidence that AI is entering medicine with the same caution as any new therapy. What the volume actually demonstrates is a different question.

Loganathan and colleagues went through the FDA's own list device by device, with six reviewers abstracting the premarket summaries and resolving disagreements by consensus. These were the same 1,016 devices Singh and colleagues classified by clinical function. Of those, 96.5% reached the market through the 510(k) pathway, 3% through De Novo, and fewer than 1% through any premarket approval. 99% of current clinical AI hit the market without any study at all (Singh et al., 2025; Loganathan et al., 2026).

The 510(k) pathway asks whether a device is substantially equivalent to something already cleared (its predicate). It does not ask for a trial. To be "like" something requires knowing what is in it: only 55.8% of the devices disclosed anything about the algorithm beyond generic language, and 15.6% disclosed nothing specific at all. In a separate analysis of predicate chains (the product the AI is "like"), Muehlematter and colleagues found that for 32.6% of devices, the very first predicate in the chain contained no artificial intelligence whatsoever (Muehlematter et al., 2023).

An AI-enabled device can be cleared on the grounds that it is substantially equivalent to a device with no AI model at all. These are devices that reach directly into patient care. It would be equivalent to a pharmaceutical company saying a drug needs no trials because it does roughly the same thing as an older drug, even though it contains a brand-new component never studied.

The FDA label only tells you the agent was permitted, not whether it works, in your building, in your patients, in the hands of your clinicians. Those are separate questions, and clearance answers none of them.

Most vendors now operate behind that proprietary boundary. They cannot tell you what is in the model for business reasons and often do not know even after digging. Some AI models even modify themselves after deployment, so the product you validated in March is not necessarily the product running in September. And, there is no post-market surveillance for these products remotely comparable to what a drug of similar reach would warrant.

This is a serious accountability gap. The ethical literature on clinical machine learning saw it coming before most health systems did: transparency, fairness, and who answers when the model is wrong do not resolve themselves once a product ships (Char et al., 2018). This is the environment the Diagnostician works in today, and naming it accurately is the first step of the intake.

The wave of exuberant clinicians and vendors keeps growing.

WHAT IS IN THE MEDICINE CABINET

An executive trying to govern clinical AI needs it organized the way a physician organizes a pharmacopeia, by mechanism rather than by vendor. Six classes are worth telling apart.

Diagnostic models aimed at detecting specific aspects of disease are the most mature class, dominated by imaging: detecting nodules, hemorrhage, breast lesions, retinopathy, and arrhythmia at or beyond the limit of the unaided eye.

Predictive and risk-stratification models forecast future trajectory from the patient's data. Things like deterioration, readmission, sepsis, length of stay, and denials are far more variable in quality and far more sensitive to local population than the diagnostic class.

Decision support models using AI for the first time to guide decision-making are being woven into the alert layer described in Chapter 12, with the complication that the rules are no longer written, or even understood, by a human.

Ambient documentation captures the encounter as an audio file and returns an increasingly structured note.

Generative models produce text, summaries, and recommendations from natural language, and are being pointed at inbox triage, discharge summaries, prior authorization, coding, and a hundred other operational uses.

Agentic systems take multi-step action on behalf of a clinician or a patient. Activities like scheduling, messaging, coding, or coordinating care.

Each class carries its own indication, its own contraindication, its own risk, and its own adverse effect profile. The last is different from the rest. Up to now, when a system recommends a course of action in a patient-facing situation, a human clinician still decides and takes accountability. We are underprepared for the interface that follows when a system acts on its own. The threshold for oversight is not incrementally higher, it is categorically higher, and no organization has built for it.

THE FIRST EFFICACY DATA

When governing AI, a Diagnostician wants evidence of impact on the illness. For most of the last 3 years the evidence for clinical AI has been demonstrations and testimonials. That has changed, and the change is recent enough that most boards have not seen it yet.

Start with burnout, because that is what ambient documentation was sold on. Olson and colleagues ran a quality improvement study across six health systems, following 263 physicians and advanced practice clinicians for 30 days after they were given an ambient scribe. Among those in ambulatory clinics, burnout fell from 51.9% to 38.8% (Olson et al., 2025). Cognitive task load improved, after-hours documentation fell, and attention to the patient during the visit improved.

That is a real effect, and it is the largest single intervention against record-driven burnout to date.

The scale is also large. Tierney and colleagues reported on the deployment across The Permanente Medical Group in Northern California, where 7,260 physicians used an ambient scribe in more than 2.5 million encounters over 14 months, with roughly 16,000 hours of documentation time returned (Tierney et al., 2025b). That is not a pilot, that is a drug in widespread use.

Now the money, because this is the book that counts it. Holmgren and colleagues at the University of California San Francisco did what had never been done before: they took 1.2 million ambulatory encounters across 1,565 attending physicians, 698 of whom had adopted an ambient scribe, and ran a difference-in-differences analysis on what got billed (Holmgren et al., 2026).

AI adopters generated 1.81 more relative value units per week than non-adopters and completed 0.8 more encounters per week, with no increase in claim denials. At the 2025 Medicare fee schedule, the authors converted that figure themselves: 1.81 relative value units a week is $3,044 per physician per year. At 3%, the post-margin benefit was about $91.32 per year.

That is the first hard financial return figure for the flagship clinical AI product in American healthcare, from a well-designed study at a serious academic center. It is not enough of a benefit to pay for the license.

The authors couldn't say whether the extra billing reflects more clinical service or more complete coding of service.

None of this is AI bashing. It just means the agent has been deployed for 3 years and we are only now measuring it, a sentence that should not be true about a technology this widespread in patient-facing healthcare.

THE UNRECORDED ADVERSE EVENTS

Every agent has an adverse effect profile. AI has four that are documented. What is the risk of bringing AI into the clinic?

Fabrication and Hallucination

This is the generation of false or invented information. In the most careful accounting available, Asgari and colleagues had clinicians annotate 12,999 generated note sentences across eighteen experimental configurations and found a hallucination rate of 1.47% and an omission rate of 3.45%. Of the hallucinations, 44% were graded as major (Asgari et al., 2025).

Roughly 1 AI-generated sentence in 70 is invented, and almost 1 in 2 of the invented sentences matter clinically. In a system generating 2.5 million notes, that is not insignificant.

Then the same team did something else: by revising the prompts and the surrounding workflow, they drove the major error rate below the published error rate for notes written by humans. It turns out that the fabrication rate is not a fixed property of the AI model; it is a property of how the model was configured by the local users. It is a governance variable, currently left at the vendor default across most of American healthcare.

Propagation

This is erroneous AI-generated content spreading into other content. The early evidence is small but pointed. Rabotin and colleagues scripted 5 interpreter-mediated Spanish and English encounters containing 20 deliberate interpreter errors and ran them through 2 commercial ambient scribes. Both scribes wrote the interpreter's errors into the note, one in 55% of cases and the other in 60% (Rabotin et al., 2026).

Errors originating in the physician's speech were propagated between 20% and 30% of the time. Errors originating in the patient's speech were propagated between 80% and 100% of the time.

This is a small, simulated study with scripted errors and two vendors, and it needs replication before policy is built on the exact figures. I am citing it anyway, because the direction is not subtle: the system weighted the physician's words and discounted the patient's, especially in encounters involving the 26 million Americans who prefer a language other than English.

Underperformance

This is when the model does not do what the vendor said. Wong and colleagues externally validated a proprietary sepsis prediction model deployed at hundreds of American hospitals, across 27,697 patients and 38,455 hospitalizations. The vendor reported discrimination between 0.76 and 0.83. Measured, it was 0.63. The model fired on 18% of all hospitalizations and failed to identify 1,709 of 2,552 patients with sepsis, which is 2 in 3 of them (Wong et al., 2021).

Hundreds of hospitals had that model turned on. Not one of them had run that analysis.

Interruption

The fourth adverse effect is interruption, introduced in Chapter 12. Every interruption and correction spends a fraction of a finite clinical day. AI can add to that number as easily as subtract from it, unless it receives the oversight it deserves.

IATROGENIC CLINICIAN ATROPHY

One forward-looking concern about clinical AI is not what it does to the patient today, it is what it does to the clinician over time.

Every powerful cognitive aid seems to carry a deskilling risk, and this is not a new observation; it parallels my "physician pacification syndrome." Goddard and colleagues showed in a controlled prescribing study that when decision support was wrong, clinicians switched from a correct answer to an incorrect one in 5.2% of cases, purely because the machine said so (Goddard et al., 2014). Lyell and Coiera reviewed the automation bias literature and found that clinician susceptibility to bias errors tracks to cognitive load, meaning the bias is worst exactly when the clinician is busiest (Lyell and Coiera, 2017).

That literature is decades old, and it was theoretical about AI. It is not theoretical anymore.

Budzyń and colleagues studied 19 experienced endoscopists at 4 centers in Poland, each of whom had performed more than 2,000 colonoscopies. The centers introduced computer-aided polyp detection at the end of 2021. The investigators compared the quality of the endoscopists' unassisted colonoscopies in the 3 months before the tool arrived against their unassisted colonoscopies in the 3 months after. Adenoma detection fell from 28.4% to 22.4%, an absolute decline of 6.0% (Budzyń et al., 2025).

This is not AI performing poorly. This is physicians performing worse without it, after only 3 months of having it. Nineteen experts, thousands of procedures each; and their unaided skill moved in a quarter. It is the first real-world evidence of automation-induced deskilling tied to a patient outcome in medicine, and it arrived in a specialty where the outcome is a missed cancer.

The hospitalist whose differential diagnosis is increasingly shaped by a ranked list may lose some of the habits that used to build the list. The resident trained from the first day with AI backup may never build the unaided reasoning the training assumes.

The diagnostic question to put to any AI deployment is whether the tool is making this physician better or slowly standing in for them. The answer determines whether the ten-year effect is healing or atrophy, and no vendor will answer it for you.

WHO IS HOLDING THE PEN

The workforce conversation is usually framed as binary: either the machines replace clinicians, or the machines return clinicians to the work they trained for. Neither framing is adequate. The better answer may be "it depends."

The composition of the work is changing. Documentation overhead is falling for clinicians who use these tools well. Low-acuity inbox volume is being redirected, some to the AI model and some back onto less expensive medical assistants and triage nurses; a transfer rather than a fix. In coding, scheduling, prior authorization, and billing, generative systems are quietly absorbing large fractions of what was exclusively human cognitive labor three years ago.

Organizations handling that well are redesigning the work around the technology. Organizations handling it badly are reducing headcount on the strength of a vendor's projection, and are about to watch a generation of institutional knowledge walk out the door. They are the only people in the building who knew why the workflow was built that way.

The related risk is mediated communication: portal triage, summarized handoffs, generated discharge instructions, translated patient education, coordinated follow-up. Each has value. Each, configured carelessly, inserts a layer between people who used to speak to each other directly.

The risk is not that the system will say the wrong thing, although it sometimes will. The risk is that the clinician stops noticing what is being said on their behalf, and the patient receives something faster and smoother and slightly less true to their situation than what a person would have said.

Deciding the workforce strategy is not the clinical informatics leader's job. Being the clinical voice in the room when it is decided is exactly the job.

THE AI FORMULARY REVIEW

The framework I use in operational and consulting work is borrowed deliberately from clinical pharmacy, because the analogy holds. What follows is how to do it correctly. I do not mistake correctly for easily or inexpensively, which is why it is usually not done at all.

Caution as the Formulary Standard

Nothing joins a hospital formulary without a serious review of the evidence, the benefit, the risk-and-side-effect profile, the cost against that benefit, and the payer position. Nothing reaches the patient until those answers are satisfactory. Some decision-maker is accountable. Some processes are deployed to monitor. There is a way to take it back off the patient. New AI deserves at least that, and probably more, because unlike a drug it can change after you approve it.

The governance body that closes this list does not have to be a new invention, it is Chapter 7's multidisciplinary conference. The same defined chair holds it. Every discipline is still required to show up and speak. Evidence gets examined. One plan gets signed and reviewed at the next meeting instead of left to drift. AI does not need a new species of oversight.

Concretely, a governed deployment looks like this: a named indication, written down, saying what clinical problem this tool is for and what it is not for. Local validation against the actual population before go-live, with subgroup performance reported. A monitoring plan with defined metrics and a defined review cadence. A named physician owner. Discontinuation criteria agreed in advance. A governance body with a physician chair and the authority to suspend.

Ambient AI and the Chart

For ambient documentation specifically, four more things apply. Require review and signature before any generated note is filed. Train explicitly that the output is a draft, not a product. Audit a monthly sample of notes against the encounter for fabrication and omission, and feed what you find back to both clinician education and the vendor. Test the tool in interpreter-mediated encounters before deploying it in them.

User Results May Vary

One more thing belongs here, and it is not usually treated as a governance question.

Output from these systems is powerfully user dependent. The clinician who commands the domain, reasons ahead of the AI model, and interrogates the answer with ordinary clinical skepticism will consistently outperform the clinician who accepts the first response. Iterative, specific, evidence-informed questioning is not a technical trick, it is a clinical skill, and it goes mostly untaught. That is a training gap, and it belongs to the same leader we have been discussing.

If this class of agent came with a package insert, it would read the way other package inserts read: indication, measured effect, adverse reactions, and the conditions under which you stop. That insert has not been written, so I have assembled one from the published literature. Every figure on it is published finding that no organization is required to read. And now they are at least together.

DISCONTINUATION CRITERIA

Chapter 12 argued that the ability to stop a thing is what separates a governed program from an accumulating one. The argument is the same here and the stakes are higher, because these agents act.

Before you purchase AI, the training and local data must first be monitored. Demand a 3-month evaluation and numerically measure the outcome, not just the reaction; this is almost never done. After it is live, do the same thing periodically.

Discontinue any AI model whose local validation cannot be produced. If no demonstration exists of how it performs on your patients, it has not been validated, it has just been installed.

Discontinue any AI model whose subgroup performance has never been examined. Obermeyer's algorithm passed every test that was run.

Discontinue any AI model whose measured performance falls materially below what the vendor represented, and put the finding in writing to the vendor. Wong's paper exists because one academic center did this once.

Discontinue any ambient tool whose audited fabrication rate is not falling, or whose fabrications are not being caught before signature.

Discontinue any agentic deployment that cannot show you the audit trail of what it did and on whose authority.

Discontinue anything not defended by name. The test from the last chapter transfers intact: put the tool on a screen in a governance meeting and ask who owns it. Silence means it has no clinical sponsor.

THE DISCHARGE SUMMARY

This is where I separate what is measured from what is hoped.

The half-day that doesn't count. This is the largest and best-supported line. Olson measured burnout falling by more than a 25% in 30 days. Tierney measured roughly 16,000 hours returned across a single medical group in 14 months. This is significant mitigation, not yet a cure; documentation burden shrinks, it does not disappear.

Money earned and not collected. Holmgren measured productivity and billing impact at $3,044 per physician per year with no rise in denials, the only rigorous figure available. It is mitigation that does not outshine the cost.

Losing the people. If the burnout effect holds, this touches the most expensive line in Chapter 8's ledger by a mechanism nothing else in the formulary reaches. No cohort has been followed long enough to know.

Harm, and what follows it. This line runs in both directions, and that is the honest statement. Diagnostic models find things the eye misses. Fabricated notes and undetected deskilling create harm that did not previously exist. Which direction your organization moves on this line is a governance outcome, not a technology outcome.

Paperwork that wastes time. Generative systems are already absorbing prior authorization drafting, coding, and correspondence at scale. This is the largest theoretical opportunity in the entire ledger and the one with the least published evidence. I am flagging it rather than counting it.

Every other agent in this formulary costs almost nothing to govern, because the asset is already bought. This one is different: it costs real money, arrives faster than the evidence, changes after you approve it, and is being purchased by people who only saw the demonstration.

Chapter 8 argued that the clinical informatics Function pays for itself many times over on the strength of reducing loss in assets already owned. Here the primary return is how you avoid buying a very expensive adverse event.

This is the place in the book where the Diagnostician earns the position by saying no, and where an organization with no Diagnostician finds out what that costs.

BEDSIDE QUESTIONS: PRESCRIBING AND DOSING YOUR AI PROGRAM

1. Do you have a chartered AI governance body with a physician chair and the authority to suspend a deployment?
2. For each AI tool running in your organization today, what is the indication, the local validation, the monitoring plan, and the discontinuation criteria on one page?
3. Has any AI model in your building been validated against your own patients rather than the vendor's? Which one, and when?
4. Has your subgroup performance ever been examined by race, ethnicity, language, and payer? If it has not, you do not know whether you are running Obermeyer's algorithm.
5. If you have deployed ambient documentation, what is your audited fabrication rate, and who audits it? Have you tested it in an interpreter-mediated encounter?
6. Who is accountable for the question of whether your deployments are building clinical skill or eroding it? If the answer is elusive, note that the first published deskilling signal appeared in three months.
7. Has your AI strategy been reconciled with your workforce strategy and your training pipeline, or are the three being managed in separate rooms?
8. When did you last suspend an AI deployment because the data did not support continuing? If the answer is never, your governance is theoretical.

Any gaps in these answers are the gaps in AI governance this class of agent requires and does not have.

. . .

ARTIFICIAL INTELLIGENCE IS THE MOST CONSEQUENTIAL agent in the modern formulary and the least governed. All signs point to a powerful new therapy. An organization deploying it without clinical oversight is running an uncontrolled trial on its own patients, with no review board, no monitoring protocol, no stopping rule, and no accountable physician. Let us now evaluate evolving care at a distance.

CHAPTER 14

TELEHEALTH AND DIGITAL HEALTH

THE ROUTE OF ADMINISTRATION

Telehealth is not different medicine. It is bedside medicine practiced at a distance. The organizations that have built programs that work understood from the beginning that the bedside discipline must move along with the bandwidth.

A patient with a chronic inflammatory condition has been on the same medication for years. It started as an infusion, a few hours in a chair at the clinic every several weeks, an appointment built around driving there, waiting, and driving home. A year ago, the same drug became available as a shot the patient could give at home.

No more clinic visits, though the patient never got used to the needle and dreaded it every time.

Now the drug came as a pill. The patient switches without hesitation, relieved to trade the injection for something swallowed at breakfast.

The infusion, the injection, and the pill are the same medication, working the same way once they reach the bloodstream. What differs is how each patient gets there. Some patients need the closer monitoring an infusion chair provides. Others manage a stable disease perfectly well with a shot at home; others would simply rather swallow a pill. None of the three is a lesser version of the drug. Each is a tool in the same arsenal, chosen for the patient in front of you.

A NEW ROUTE, NOT A NEW DRUG

In the spring of 2020, American healthcare was pushed overnight into an experiment that no organization would have designed on purpose, and a generation of clinicians and operators learned something new.

Telehealth, properly configured, was not a bad substitute. For a defined subset of patients and conditions it performed as well as the alternative. For patients who had struggled for years with transportation, childcare, mobility, or geography, it was the first form of care that ever reliably met them where they were. The access gain was not evenly distributed: older patients, Black and Hispanic patients, patients who preferred a language other than English, and lower-income patients used it less, and used video less than others (Eberly et al., 2020).

Six years later it has settled into something neither the enthusiasts nor the skeptics predicted. It did not replace in-person care. It did not collapse. It became a modality of healthcare delivery.

In pharmacology, the route of administration is not just a detail. Oral, intravenous, intramuscular, inhaled, and topical are the same drug molecules delivered differently, and the differences are consequential. The route can affect the uptake, how the liver processes it, and how long it takes to act; what can be given by one route cannot always be given by another. The oral form is not a lesser drug, it is a different route with its own indications.

That is the correct way to think about telehealth, and it has eluded healthcare leadership, which tends to think of it as a product line.

A product line has a revenue target. This is the same service delivered by a different route.

GAINING THE HOME ENCOUNTER

Before we establish the indications, we should discuss what transmits easily across the "distance" medium and what does not. History transmits almost completely. So does the medication review, the trend, the shared decision, the teaching, the plan, and the relationship, if one already exists. Most of what a physician does in a room is conversation, and conversation is what this medium was built for.

Examination transmits only partially, unevenly, and sometimes not at all. Gait, affect, respiratory effort, and the general impression of a patient who looks unwell survive video visits to some degree. Touch and palpation do not. Heart and lung sounds do not, without a special device. And, the low-grade sense experienced clinicians rely on most, that this patient is sicker than the numbers say, arrives unreliably. There is no way to tell by how much.

What the remote medium adds is often discounted. You see the home. You see whether there is food, who else is in the room, whether the medications sit on the counter, whether the oxygen tubing runs where it should. Some readmissions are explained in a single video call that would never have been explained in clinic.

So the route is not a lesser version of the bedside. It is worse in some areas and better in others, and a clinic that has not worked this out is struggling without knowing why.

WRITING DOWN THE INDICATION

The indications for telehealth cluster into five groups, and each is strongest where the clinical content of the encounter is mostly conversation.

Continuity for established patients with stable chronic disease, the well-controlled diabetic, the hypertensive on a settled regimen, the patient on maintenance psychiatric medication. For these patients the virtual visit is frequently better than the office visit; it is more accessible, less disruptive, and far more likely to occur.

Mental and behavioral health. Across psychotherapy for depression, anxiety, and post-traumatic stress, the accumulated evidence shows video delivery performing at or near in-person delivery, with better access and lower missed-visit rates. In many systems it has become the modality of choice rather than the modality of last resort.

Inpatient specialty consultation. The neurology consult at a rural hospital, the complex second opinion across a region, the pediatric subspecialty visit that would otherwise be a six-hour drive. Here the alternative to telehealth is not worse care, it is no care, and the clinical benefit belongs to the route itself.

Remote monitoring and asynchronous care, covering blood pressure, weight, glucose, postoperative recovery, postpartum follow-up, secure messaging, and store-and-forward images. This category holds the most evidence and, as the next section shows, the widest range of performance.

Acute care extension. Tele-critical care, tele-stroke, virtual nursing, emergency psychiatric consultation, remote expertise pushed into a setting that lacks it. It is the oldest use of the route and still one of the best supported.

WRITING DOWN THE CONTRAINDICATION

Telehealth is a poor route for *first-encounter diagnostic work* that needs an examination. The failure mode is specific, the physician substitutes the history for a detailed examination, the differential narrows on incomplete data, and the miss is only discovered when the patient turns up somewhere else.

It is a poor route for *high-acuity presentations needing immediate intervention*, where every minute of remote conversation is a delay a bedside encounter would not create. It is a poor route for *relationships that do not yet exist*; the satisfaction evidence is consistent here: patients value virtual care with a physician that they know and then dislike it with one they do not. It is a poor route *wherever the absence of bedside cues is itself the risk*, the patient who looks worse than they sound, an impression no monitor delivers. And, it is a poor route *where connectivity is unreliable*.

An organization that has not written its contraindications down has only addressed telehealth capacity.

THE ACTIVE INGREDIENTS

Start with the intensive care unit, where the stakes are highest and the evidence is best. Kalvelage and colleagues pooled 20 studies covering 477,637 patients, asking not whether tele-critical care works but under what conditions. They separated the studies by how much authority the remote team held. Where the remote intensivist team had decision-making authority, intensive care unit (ICU) mortality fell, and both intensive care and hospital length of stay improved (Kalvelage et al., 2021). Where the remote team could only observe and advise, the benefit was not there.

Same cameras. Same monitors. Same intensivists. What separated a program that saved lives from one that did not was whether the authority to act remotely had been written down.

The same shape appears in a different environment. Tucker, McManus, and colleagues assembled individual patient data from randomized trials of blood pressure self-monitoring, the simplest form of remote monitoring there is. Overall, self-monitoring lowered clinic systolic pressure by about 3.2 mmHg at 12 months against usual care.

Self-monitoring alone, with no response on the other end, moved systolic pressure by 1.0 mmHg. Combined with intensive support, meaning a team was reliably receiving the readings and adjusting therapy, it moved 6.1 mmHg (Tucker et al., 2017).

The device is identical in both arms. The measurement is identical. What differs is whether a human being was accountable for doing something when the number came in, a sixfold difference.

A third useful example. Li, Ngai, and Genes examined 257 telemedicine encounters for acute respiratory infection, comparing prescribing between physicians affiliated with the health system and physicians supplied by an outside vendor working the same platform for the same patients. Affiliated physicians prescribed antibiotics in 18% of encounters. Vendor physicians prescribed in 37% (Li et al., 2024).

That pattern holds at scale. Ray and colleagues compared antibiotic prescribing for children with respiratory infections across three settings and found antibiotics prescribed at 52% of direct-to-consumer telemedicine visits, 42% of urgent care visits, and 31% of primary care visits, with guideline concordance lowest in the direct-to-consumer telemedicine group (Ray et al., 2019).

The explanation is not technical. The camera was the same. The patients were comparable. One group of physicians worked inside a clinical governance structure that measured them, and the other did not.

Tele-critical care saves lives when the remote physician is authorized to act. Remote monitoring works 6 times better when there is accountability for the reading. Virtual visits prescribe half as many unnecessary antibiotics when the prescriber sits inside your governance rather than a vendor's.

Three agents. Three literatures. Three different sets of investigators who have never cited one another. One finding.

The technology is almost never the variable. The variable is whether a physician with authority stood behind it, and that is the entire thesis of this book from three different angles.

THE SAME MEDICINE OR A NEW SERVICE LINE

The characteristic mistake of early clinical programs was building telehealth as a separate product. Separate documentation standards. Separate billing. Separate oversight, quality measurement, and governance.

The result was a parallel system that performed worse than the in-person alternative, and not because the medicine was worse. The same clinical discipline had not been carried across both.

Mature programs make the opposite choice. The same documentation standard. The same quality measurement. The same decision support firing in the virtual interface that fires in the exam room. The same safety event reporting, coding and compliance discipline, care coordination, and accountability.

The route changes. The medicine does not.

THAT SENTENCE IS A CLINICAL INFORMATICS PROGRAM AT WORK. Structured data must come out of the virtual encounter and land where quality measurement and revenue cycle can read them. Decision support must deploy into the telehealth encounter just as it does in the clinic. The record must be integrated so a clinician seeing the patient in person can read a previous virtual note without hunting for it. The patient experience must hold together across the organization.

None of that happens because a platform was purchased. All of it requires a Diagnostician who understands both the modality and the underlying infrastructure, and who governs the seam between them.

The seam between in-person and remote visits is where the failures live. It is reliably neglected, and the part usually never owned by an assigned role.

THE EXPIRING PRESCRIPTION

One structural point belongs in front of any board approving telehealth investment, and I have never seen it presented there.

The entire American telehealth expansion rests on Medicare flexibilities created as temporary pandemic measures and extended by Congress repeatedly, sometimes with days to spare, occasionally lapsing in the gap. The current extension runs only through the end of December 2027 (Consolidated Appropriations Act, 2026).

The geographic restrictions, the ability to treat a patient in their own home, the audio-only allowance, the ability of federally qualified health centers and rural health clinics to serve as distant sites, all of them operate on that clock.

I am not predicting what Congress will do. I am pointing out that a modality supporting rural access, behavioral health capacity, and a meaningful share of chronic disease management for the country's largest payer currently exists on a renewable lease. Most strategic plans I have read treat it as permanent infrastructure.

The clinical informatics leader is the person in the room who should know the expiration date, and signal what happens to access, staffing, and margin if it is not reauthorized.

DISCONTINUATION CRITERIA

The formulary discipline holds here too. Telehealth programs almost never get stopped.

Discontinue any telehealth service line that has no written indication. If no clinical problem is named, it will drift.

Discontinue any service line whose documentation, quality measurement, or safety reporting runs on a separate track from its in-person equivalent. That is a parallel system, and it produces outcomes you cannot see.

Discontinue any arrangement in which clinicians treating your patients sit outside your clinical governance, unmeasured on your terms. The prescribing data are unambiguous about what that costs.

Discontinue any remote monitoring program with no accountability for acting on the reading within a defined window. An unmonitored monitor is a liability.

WHAT TELEHEALTH RETURNS

Losing the people. Telehealth capability is what allows a rural hospital to keep a service line, a health system to cover a specialty it cannot recruit for locally, and a physician with caregiving obligations to stay in practice at all. This is a retention and coverage mechanism more than a revenue mechanism, and the organizations that understand it that way get the most from it.

Care we did not need to give. This depends entirely on governance. Well-governed virtual care substitutes for a visit. Ungoverned direct-to-consumer virtual care adds a visit, an antibiotic, and sometimes a follow-up in person for the same complaint.

Harm, and what follows it. Tele-critical care with decision-making authority reduces intensive care mortality. Tele-stroke gets thrombolysis to patients who would not otherwise be evaluated in time. Against that, missed diagnoses from examinations that never happened, in a category poorly counted, and readmissions never attributed to the virtual visit.

Paperwork that wastes time. Negative wherever a parallel system has been built. A program with separate documentation and separate quality abstraction has added administrative work, not removed it.

The price of admission is low. Minus the cameras, the platform is already bought. The clinicians are already employed. The patients are already asking. What is missing is a written indication, a written contraindication, an integration plan, and a clinician who is allowed to act at a distance.

BEDSIDE QUESTIONS: PRESCRIBING AND DOSING YOUR TELEHEALTH PROGRAM

1. Can you produce your written telehealth indications, and are they operationalized in scheduling and intake, or do they exist only in a policy document?
2. Can you produce your written contraindications and your escalation pathway from virtual to in-person? If not, your clinicians are each deciding privately.
3. Are your virtual encounters documented, measured, and safety-reported to the same standard as your in-person encounters, or on a parallel track?
4. Does your decision support fire in the telehealth interface? Has this been checked, rather than assumed?
5. In your tele-critical care, tele-stroke, or virtual nursing programs, does the remote physician have decision-making authority in writing? This is the single question in this chapter most associated with whether the program saves lives.
6. In every remote monitoring program you run, who receives the reading, within what window, and with what authority to change therapy? Name them.
7. Are any clinicians treating your patients virtually outside your clinical governance? How does their prescribing compare with your own? If you have never compared, the published gap is roughly twofold.
8. Can you report telehealth utilization by race, ethnicity, language, age, and payer, and can you report video share separately from total telehealth share?
9. Who in your executive group is accountable for the integrated patient experience across modalities? If there is no answer, your program is producing outcomes you cannot see.

10. What is your organization's exposure if the current Medicare telehealth flexibilities are not extended past their expiration? Has it been modeled?

Ensure that your answers don't reveal a telehealth program running as a parallel system that does not hold together across settings, rather than an integrated route of administration.

THE ARMORY IS NOW CHARACTERIZED. THE RECORD AS THE *imaging study. Decision support as a first-line agent. Artificial intelligence as the newest medication. Telehealth as a new route of administration. Four instruments, each with an indication, a dose, a monitoring requirement, and a criterion for stopping.*

The treatment has been prescribed. Is it working? How do you know? Who is watching? That discipline, the clinical method practiced as continuous organizational habit rather than as a project, is Part VI.

PART VI

MONITORING THE RESPONSE

The treatment has been initiated. The patient's response, whether favorable, partial, paradoxical, or absent, must now be read, interpreted, and acted on. Let us examine the clinical method as it applies to the ongoing management; the disciplined cycle of follow-up, reassessment, titration, and adjustment that separates a one-time intervention from durable healing.

The clinical method is the single variable that separates organizations that hold their gains from organizations that spend every 3 years redoing the same work.

CHAPTER 15
THE CLINICAL METHOD APPLIED
CARING FOR A PATIENT

The clinical method is one of the most rigorously developed reasoning disciplines in modern professional life. The physician who applies it to a patient is providing a reasoning platform that guides their patient to recovery. The physician who applies it to an organizational problem is not borrowing a metaphor. They are providing the same reasoning that produces the modern medical diagnosis to a different patient.

The patient is 26 years old and describes pain that started around the navel two days ago and has since moved down and to the right. The resident taking the history does not accept the word "pain" as an answer. Where exactly? When did it start? What makes it worse? Did it move before or after the nausea began? Each answer narrows something and widens something else, and by the time the resident's hands are on the abdomen, there is already a short list forming that did not exist twenty minutes earlier: appendicitis, an ovarian cyst, a stomach virus that will resolve on its own, or a urinary infection masquerading as something worse.

The resident does not order every test that could touch that list, only the few that will move it. A white count, high. A urinalysis, clean. The migration of the pain from the navel to the right lower quadrant is not just a symptom; it is data, a pattern more common in one diagnosis on the list than in any of the others, and the resident writes it down that way. The probability shifts by exam findings, then by labs. A quick abdominal ultrasound gives the final key measurement. Appendicitis is confirmed. The plan goes into the chart in writing, nothing by mouth, surgical consult placed, repeat abdominal exam in 2 hours, call if the fever climbs or the pain spreads before then.

The patient comes in describing pain. The resident leaves having found a disease; the difference between them is where the entire discipline lives.

THE CLINICAL METHOD EXAMINED

The discipline a third-year medical student begins to absorb on the wards is not, mainly, a discipline of facts. It is a discipline of reasoning under uncertainty, and it is the formal training most nonclinical professions never receive.

The student learns to take a thorough history, examine with eyes and other senses, generate a differential, order only the tests that discriminate, weigh the result against prior probability, narrow the list, commit in writing, prescribe with explicit monitoring parameters, and re-examine for response, all while separating the complaint the patient came in with from the disease the patient has.

That discipline has been refined over centuries, taught explicitly at every accredited medical school, reinforced through residency and fellowship, and practiced daily by roughly 1 million American physicians. With very small adaptations it is exactly what a healthcare organization needs to apply to itself, and almost no organization has the trained talent at the table.

The clinical method works, in my own experience as physician and patient, and through extensive, validated medical literature (Sackett et al., 1996). I am not claiming that American healthcare currently practices it well. I hope this book calls the profession back to it. (See Figure 15.1.)

THE CLINICAL METHOD AS DISCIPLINE

The Diagnostician's Clinical Method

A cycle, not a line.

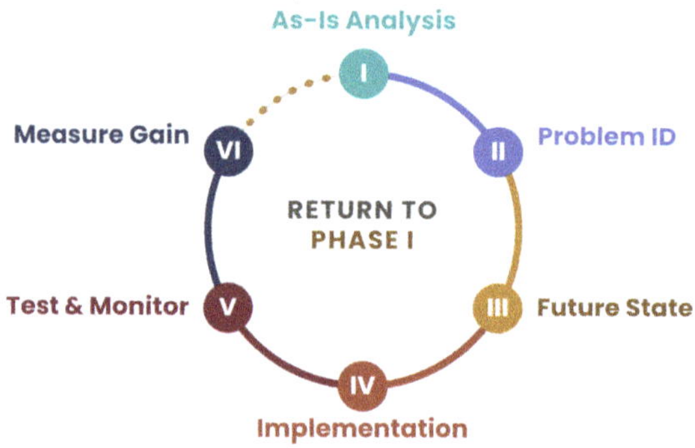

I	**As-Is Analysis** CURRENT STATE	Map the workflow as it actually unfolds. Pull the baseline metrics before touching anything.
II	**Problem ID** NAME THE DISEASE	Name the problem, and trace it back to its mechanism.
III	**Future State** WRITE THE PLAN	Specify the to-be workflow: the measures, the owners, the deltas.
IV	**Implementation** DELIVER ITERATIVELY	Execute the build. Pilot first, then generalize.
V	**Test & Monitor** MEASURE RESPONSE	Validate against the measures and the safety signals.
VI	**Measure Gain** CONSOLIDATE	Close the loop. Publish the result, hold the gain, standardize the change.

Phase VI returns to Phase I. The method is a cycle, not a line.

FIGURE 15.1

The six-phase clinical method, applied to informatics.

MOST CLINICIANS WHO HAVE WORKED ON A MODERN WARD KNOW WHAT has happened to the method. The history is taken through a shared template that is touched by too many hands, and some of it is imported from places we do not know. The physical exam is shortened for time, traded for quicker heuristics. The differential is skipped entirely, because the order sets are built for presenting complaints in a way that favors speed over reasoning. As with the AI failures examined earlier, treatment starts before the organization knows what they are aiming at. Premature closure delivered before other options are considered. The commitment to a target diagnosis reduced to a neglected problem list. Longitudinal tracking done as a team sport in an unreadable record. And, reassessment, the safety net that catches a wrong diagnosis, is the first thing dropped when the schedule is full.

Most physicians reading this will recognize each of these.

A fair reader might ask why they should accept a method from a profession visibly struggling to apply it.

THE STRUGGLE IS NOT EVIDENCE AGAINST THE METHOD, *IT IS EVIDENCE that the method has been running in an environment not engineered for it.*

That distinction is the whole point of the modern patient safety movement. The safety literature most commonly finds that error originates in the system rather than the individual. It appears in conditions that were designed rather than in the person at the bedside when it went wrong (Reason, 2000). That finding is why this book spends fourteen chapters on the environment before it spends one on the practice.

A systems explanation is an explanation. It is not an absolution. The environment is the main cause, but the practice is still ours to own.

The argument runs in both directions. An organization that installs the clinical method as an executive tool gets a better organization. A physician who applies it at the bedside gets better medicine. Bedside care is not this book's primary diagnosis, but I suspect it is a downstream casualty of the same disease.

EXAMINING AN ORGANIZATION AS A PATIENT

The history and physical is the foundation. Done well, it makes everything after it possible. Done poorly, everything that follows is a guess.

The bedside discipline is to gather relevant information before hypothesizing. The young physician hears the chief complaint, commits immediately to a conclusion, and reads every subsequent finding through that lens. That is premature closure, among the best-characterized failure modes in clinical reasoning (Croskerry, 2002). It is not a rare error, and it is not confined to trainees. Saposnik and colleagues reviewed 20 studies covering 6,810 physicians and found overconfidence, low tolerance for risk, anchoring, and availability bias associated with diagnostic inaccuracy in between 36.5% and 77% of case scenarios (Saposnik et al., 2016).

Those are trained Diagnosticians, working inside a profession that teaches this discipline explicitly, and they still close early most of the time under normal conditions. An executive team that has never been taught the failure mode at all is not doing better.

Applied to an organization, it looks like the following. You are asked to fix a stalled decision support improvement project. The temptation is to diagnose it in the first meeting. The clinical committee is unfunded, unsupported, or in over their heads. The analyst team is too small. The backlog of work is being mismanaged.

Resist it. Walk the workflow. Read the notes. Pull the override data yourself. Sit down with the users. Read committee minutes. Map who the decision-maker is. Build the entire picture from the facts before committing to a theory.

The instruments are ordinary, but the rigor is not. Do the process mapping that separates the theoretical workflow, as written in policy, from the actual workflow, as practiced every day. Direct observation with a notebook and a clock, at the moments where the work succeeds or breaks. Structured interviews asking the people doing the work about failures they have already identified and have yet to be collected. Quantitative review of what the workflow produces, attending to every gap between what the dashboard reports and what the work gathers, captured with a structured framework such as the updated Consolidated Framework for Implementation Research (Damschroder et al., 2022).

An organizational history and physical is not a 1-week exercise. It can take weeks with a small, dedicated team, and it produces a written document, long, detailed, and internally consistent, describing the patient as the flawed patient it is.

If that document does not exist, no diagnosis can be rigorously examined.

NAMING WHAT YOU ARE TREATING

The differential diagnosis is the cognitive heart of the method. It is the disciplined evaluation of everything that could account for the picture, ranked by probability and by the consequences if it were true ("What is most common?" and "What is most deadly?"), with a method that can tell them apart.

A medical student hears the words "chest pain" and jumps straight to myocardial infarction, without considering pulmonary embolism, aortic dissection, esophageal rupture, pneumothorax, costochondritis, reflux, and anxiety. They have skipped the step that would have caught the dissection. The experienced physician examines, ranks, and then orders the workup that discriminates.

The executive hears that quality scores are flat and concludes that "we need more decision support," without considering documentation failure, a workflow drift, a change in definitions, a data error at the core of the quality score, and lack of informatics and governance.

They are making the identical error in a different setting. The difference is that premature closure will never be named, because the room was never taught the term.

Commit to the working diagnosis in writing, with the evidence that supports it and the alternatives considered and rejected. That written commitment is what separates rigor from a guess. A diagnosis written down is a hypothesis the team can confirm or falsify; a guess carried in a person's head cannot even be examined. Most organizational initiatives fail this test. A project charter describes what will be done. It rarely states what is believed to be wrong.

WRITING AN ORDER THAT CAN BE EXECUTED

The treatment plan bridges the diagnosis and the work. Done well it is specific, measurable, time-bound, and owned. Done poorly it is just a list of intentions.

The bedside standard is that a colleague covering your service tomorrow could execute it without you in the room. Reduce the hemoglobin A1c from 9.2% to under 7.0% within 6 months. Metformin 1,000 mg twice daily, dietary counseling, exercise prescription, labs at 3 and 6 months. If the target is not reached at 3 months, add a GLP-1 agonist.

Specific. Measurable. Time-bound. With the reassessment point already written.

Applied organizationally, this is among the most underdeveloped capabilities in healthcare leadership, and I say that having read a great many strategic plans. Improve quality scores. Reduce burnout. Optimize the record. Those are not plans. They cannot be clearly executed. More importantly, they cannot be falsified or ever be wrong.

A good plan looks like this. Reduce the sepsis bundle override rate from 96% to 65% over 6 months, by rebuilding the alert with composite scoring and an integrated order set, governed by the sepsis clinical committee, owned by a named physician with named analyst support, override rates and bundle compliance reviewed every 2 weeks, and with discontinuation criteria if the rebuild produces unintended consequences.

These can be executed, reviewed, and judged. At the end of two quarters, the owner can be told that they were either correct, or wrong and to change course. That is the power that makes the rest of it work.

The process improvement disciplines share this structure. "Plan, Do, Study, Act" for tight cycles. "Define, Measure, Analyze, Improve, Control" for larger transformation. Both create a cycle that keeps a promising plan from collapsing. The leader who requires an action plan that can be confirmed or falsified will produce results.

STARTING LOW AND GOING SLOW

Implementation, the "Treatment" phase, is where most plans die, clinical and organizational alike. The plan was sound, the execution wandered. The patient did not comply, or complied without responding as expected. The team failed to follow through. The unintended consequence arrived before the intended one. Attention moved to something more visible.

The bedside discipline is to start low, monitor, and titrate. Give the lowest reasonable dose. Follow the patient. Check the labs. Assess the response. Adjust. If a side effect emerges, reduce or switch. If the response is partial, intensify. The implementation of treatment is a continuous conversation between the prescription and the patient.

Organizationally, the same rules hold. The new decision support rule goes to one small group first and is monitored before it generalizes. The Ambient AI is trialed for 3 months in the local environment, with performance metrics captured and user feedback collected. The workflow change is piloted with a small cohort, with user feedback collected, before it scales. The clinical decision support (CDS) model is deployed in stages, with override rates and usability tracked, before it goes enterprise wide. Clinical initiatives are never plug-and-play. Never.

A good implementation is not a switch thrown on a single go-live date. It is a measured titration.

The leader who spends part of every week on the units, talking to the people using the new build, hears the friction the dashboard was never designed to capture. That is the evaluation of the treatment, and it cannot be delegated to a survey. It is also the hardest part of the job.

THE UNSCHEDULED FOLLOW-UP VISIT

Follow-up is the phase most often skipped, and skipped for the same reason in both settings. The acute work is finished, attention has moved, and looking back is never rewarded.

The patient starts the medication and goes home. At some point they return, and by then the original treatment has worked, has not worked, or has produced an adverse effect that should have been caught weeks ago.

Most informatics initiatives are evaluated, if at all, after the project has closed, the team has dispersed, and the sponsor has moved to the next thing. By then course correction is not only difficult. It is structurally impossible.

The discipline is to design the follow-up into the plan before the work starts. Override rates checked at week 2, week 6, week 12, and quarterly after. Physician experience surveyed at month 4. Safety events monitored continuously. Discontinuation criteria written down in advance and triggered.

Follow-up that is written into the plan at the beginning is called *monitoring*. Follow-up that is added at the end is called an *audit*, and an audit occurs when monitoring was neglected.

HOLDING THE TREATMENT GAIN

The last phase is to capture what worked, why, and how to keep it.

Clinically, this is the conversation at the end of an acute episode. The illness is treated, the patient is recovering. Now the maintenance begins, the medications that continue are refilled, the lifestyle changes that hold the gain are logged, the surveillance that catches the next flare-up is scheduled. Recovery is not the end of the work, it is the beginning of maintenance, and every experienced physician knows a patient who did beautifully, until they didn't.

Organizationally, the intervention that lowered the override rate, increased the documentation yield, and improved the compliance number is now *a capability*. A capability must be staffed, funded, governed, onboarded for every new clinician, analyst, and leader, and measured continuously. Every gain, clinical or organizational, is vulnerable to drift when left alone over a long enough timeline.

Treat the maintenance with the rigor of the original intervention and the gain holds. If you declare victory and move on, your successor will do this same work again in 3 years as if it were new.

ONE OPERATIONAL CASE, PRESENTED IN FULL

A hospital system presents with a chief complaint of rising claim denials.

Examination. The informatics leader and the revenue cycle team walk a sample of recent denials end to end, the notes, the order sets, the structured data capture, the payer's review logic, the coders, the documentation workflow on a busy day. They produce a written summary of the workflow as it is, the data capture as it is, the payer's actual logic, and the friction points found.

Diagnosis. The differential includes coding error, insufficient documentation, structured data capture gap, payer logic shift, and workflow drift. The examination points hard at two of these and less at the other three. The working diagnosis is committed in writing:

denial volume is driven primarily by a structured data capture gap in three high-volume order sets, with a secondary contribution from documentation workflow drift in one service line.

Plan. Rebuild the three order sets to capture the elements the payer's review requires, governed jointly by clinical informatics and the affected service line. Redesign the documentation workflow for that service line so the required data are easy to enter and sit inside the standard note rather than another screen.

Target: reduce the denial volume in the three associated groupings by 40% within 4 months. Named owner, named analyst team, review every 2 weeks during the build and monthly after.

Treatment. Staged. One order set first, released with 2 weeks of monitoring before the second and third, the workflow changes piloted in one clinic before scaling, clinician feedback collected at every step.

Follow-up. The denial volume dashboard is read every 2 weeks, coder feedback solicited, payer logic changes watched by a named owner and flagged, clinician-reported usability built into the standard surveys. The plan adjusts to whatever friction surfaces.

Outcome. At four months, denial volume in the three groupings has fallen 60%, better than the target.

None of the day-to-day work was easy. None of the outcome was obvious early.

The first order set rebuild added 90 seconds of documentation time per encounter, and the service line said so within a week. That is the re-examination working exactly as designed. Physician time was a balancing measure, reviewed at the 2-week checkpoint, and the team revised the build before modifying the second and third order sets.

The denial volume did not move at all until month 2. Judged at week 3, the plan would have looked like a failure, and would have been killed.

Maintenance. Joint clinical and revenue cycle order set governance is made permanent, with a quarterly review and a published metric. The two leaders hold an after-action review and write down the lessons. The next denial pattern will be worked the same way.

None of that is exotic. It is the clinical method, applied with discipline, by people who were trained in it. It is what every executive team should be able to do and what every board should expect.

BEDSIDE QUESTIONS: THE MONITORING AND DISCIPLINE OF YOUR METHOD

1. Are your major initiatives launched with a written examination, a written diagnosis, and a written plan, or with a goal and a project charter?
2. On your three largest current initiatives, can the team hand you the working diagnosis in writing, with the alternatives that were considered and rejected? If it exists only as a thought, it cannot be checked.
3. Are your plans specific, measurable, time-bound, and owned, or are they intentions? Pick one at random and test whether a colleague could execute it without the author present.
4. Is the follow-up cadence written into your plans at the start, or added at the end as an audit?
5. Is implementation titrated, or does it go live everywhere on a date picked without clinical input?
6. After your last successful improvement, was the capability chartered, staffed, and funded, or did the team disband and the gain drift?
7. For your last three major proposals, was the breakeven point calculated or margin multiplier, from Chapter 8, applied?
8. Has your executive team ever been taught this method explicitly, or is it a random discipline that some people happen to have?

Any gaps in the answers suggest planning without diagnosing, implementing without titrating, and declaring victory without measuring.

THE CLINICAL METHOD HAS NOW BEEN TAUGHT. THE diagnosis was made. The plan was written. The treatment was titrated. The response was measured. The gain was held.

The patient is recovering, and the physician's obligation now is the honest, calibrated conversation about what happens next. And, the prognosis.

PART VII

HEALING, REHABILITATION, AND PROGNOSIS

The acute episode has been treated. The patient is recovering. The work now is to consolidate the gains, build the maintenance plan, and anticipate the comorbidities arriving on the horizon. Now the long course: the watch list of pathologies the patient is likely to be exposed to next. A prognosis is an educated prediction. It tells the executive what the next decade will require if the recovery is to be permanent. And, it ends with the referral: recognizing what no single organization can treat inside its own walls, and writing the letter that says so.

CHAPTER 16
THE PROGNOSIS
COMORBITITIES ARRIVING ON SCHEDULE

THE DISEASES OF THE NEXT DECADE IN AMERICAN HEALTHCARE ARE PARTLY familiar, many of them are old conditions in new forms, and a few will be novel. The executive-physician is the one whose work is to anticipate them, prepare for them, and treat them when they arrive. The patient is your organization, and you are preparing for the long course.

THE DOCTOR HAS NOT FINISHED SITTING DOWN BEFORE THE PATIENT asks, "What happens next?"

The physician does not offer a single number, because the honest range is wide. Some patients with this degree of weakened heart muscle recover real function over the following year. Most stabilize at something less than before, but livable. A smaller group progresses toward needing a device, or a transplant evaluation. What moves a given patient from one end of that range toward the other is well known: the medication taken every day and not skipped, the weight checked at the same hour each morning, the salt left off the plate, the follow-up visit kept. These do not guarantee an outcome; they simply change its odds.

The hard decisions cannot wait for the picture to get clearer. Whether to move now toward evaluating for an implantable defibrillator. Whether to have the conversation about what kind of care the patient wants if the treatment doesn't work. The physician lays out the prognosis. The patient can now plan.

A PROGNOSIS DONE WELL

A prognosis is not a prediction. It is a disciplined conversation, and three things separate it from a guess: a range rather than a point, because a single number misses what probability is; the modifiable factors named apart from the ones that are not, because modifiable is the only part a patient can act on; and honesty even where compassion would rather soften it, because a sanitized prognosis cannot be planned against.

This chapter is that conversation, held for the American healthcare organization.

CURRENT VITAL SIGNS

Picture a regional system, 4 hospitals, 18 ambulatory sites, 1 million covered lives, opening its dashboards on an ordinary morning. Ambient documentation has captured the overnight notes. Predictive models have flagged seven patients for early rapid response. The artificial intelligence governance committee meets at noon, and the Diagnostician is already in the room where those decisions are made.

That is not a story for 2035. It is happening now, in the organizations that began this work three to five years ago.

There are a number of looming conditions arriving on a schedule that the industry does not control. Each is described below, and we then identify the single variable that determines whether any of them can be survived.

THE COMORBIDITIES ARRIVING ON SCHEDULE

Payment for Bad Documentation

The move from fee-for-service toward value-based payment has been the slow arc of American reimbursement for two decades, accelerating without ever completing: Medicare Advantage growing as a share of Medicare, total cost of care arrangements, bundles, and accountable care organizations expanding in scope and rigor, state Medicaid programs increasingly demanding value-based contracting, and commercial payers in many markets following.

In my opinion, much of what is being imposed was never healthcare's role in patient care. Whether you think the payment trajectory is good or bad, the industry was not instrumented for it and still is not. The care team approach, longitudinal tracking, data content and quality, personnel volume, population tools, external monitoring, and critical communication and coordination; these are from some other agenda. It is not what most clinicians were trained to do. And, whether it is reasonable, successful, doable at scale, or doable at the price, like every other treatment we have discussed, will only be evaluated after the damage is done.

The consequence is an informatics one. Infrastructure adequate for fee-for-service is inadequate for risk, including: risk stratification, panel management, gap closure at the point of care, social determinants captured as data rather than as narrative, longitudinal cost tracking, and population analytics. Most EHR vendors have built tools for these requirements. Most organizations have configured pieces of them. Almost none have built them into the workflow.

The next decade is partly the decade those capabilities must become as dependable as EHR uptime. A leader who spent years on optimization, governance, and data quality is positioned for it already. One who spent the last five years reviewing vendor release notes is not, and will find out soon.

AI Multiplying Faster Than Governance

Chapter 13 examined artificial intelligence as a class of therapy. The next decade is when it stops being a special category and becomes routine: AI-assisted dictation and ambient documentation standards; predictive models aimed at the local population and generating care interventions; generative systems managing inbox triage, patient communication, prior authorization, coding, and revenue cycle functions; agentic systems taking multi-step action inside our walls on increasingly complex workflows.

I have high hopes for this category. Designed well, used responsibly, and targeted at real problems, a clinician practicing with AI enhancement has a definitive advantage over one who is not.

The governance problem, however, does not get easier. It gets larger and more dispersed. The number of AI models running in a typical health system before this decade ends will almost certainly exceed the number of decision support rules running today, and Chapter 12 already established that most organizations cannot govern the rules they have now.

Every discipline Chapter 13 described must scale: local validation before deployment; subgroup performance monitored after; fabrication audited; deskilling watched for; discontinuation criteria written in advance and triggered. None of this is optional diligence, it is the ethical floor for deploying a system that reasons about a patient, and the literature named it before most health systems even had a governance committee to enforce it (Char, Abramoff, and Feudtner, 2020). That volume of work is already a growing workforce problem.

The Organs That Cannot Be Replaced

The American clinical workforce is older, smaller, and more strained than at any point in living memory, and the projections disagree with each other.

The Association of American Medical Colleges projects a national physician shortage of between 13,500 and 86,000 by 2036 (AAMC, 2024). The Health Resources and Services Administration projects the same shortage at roughly 187,000 by 2037 (HRSA, 2024). These are two serious projections of the same shortage that differ significantly. Despite no clear point estimate, both are pointing in the same concerning direction.

The consequence for an executive is not just numeric. It is the end of every workforce assumption we have relied on.

For two decades American healthcare ran on abundant clinical labor, absorbing the inefficiencies of badly designed technology and operations. That assumption has expired. Every minute of clinical time consumed by an avoidable workflow is a minute that cannot be bought back. Every hour of after-hours documentation that moves a physician toward early retirement is an hour the organization will not be able to recruit. And, everything I have just said applies to every other clinician in the building, and increasingly to every other position supporting operations.

Technology investment over the next decade will have consequential workforce impacts. It must be judged on what it newly enables, and more, on whether it returns time to people the organization cannot replace.

It also opens the most consequential question healthcare has yet asked of itself: "*Who is replaceable?*" Environmental services, overtaken by a robot. The call center, replaced by agentic AI. The executive, by an automated algorithm. Even the physician, by AI decision support.

These are not questions to answer after deployment, not in an industry with life-or-death consequences, with no governance in place, and clinicians already working at the edge of their ability.

They need to be thought through, carefully and rigorously, in advance.

This is the most operationally consequential condition in the chapter, and the one Part V addresses most directly.

The Record That Finally Arrives

Something is happening in interoperability that has been promised for two decades and is now measurably real.

The Trusted Exchange Framework and Common Agreement (TEFCA) designated its first qualified networks at the end of 2023. Within these years, the volume of patient records moving across that framework went from under 10 million health records to more than 1 billion (HHS, 2026).

The longitudinal record fragmented across organizations is becoming whole; the growth curve matters more than the current total. The patient who has repeated their history to every new clinician they see now walks into encounters where it has already arrived.

The arriving record is worth exactly as much as the workflow that consumes it and the quality of the data it carries. A complete external history that is never read is no improvement on a fragmented one, and a consumed history with corrupted elements brings bad data into a clean encounter.

What the receiving clinician sees, when, in what context, and with what flags, is a design problem. Converting text into discrete usable elements is an informatics problem. It is one of the more interesting data challenges in the field. It will be solved well in a small number of organizations and not at all where the Diagnostician does not exist.

WHAT REMAINS FOR THE CLINICIAN

Underneath all of this sits a quiet question, and a confident answer to it is probably wrong.

What does the clinician's work look like once ambient systems have captured most of the encounter, generative systems have drafted most of the inbox, agentic systems handle much of the coordination, and predictive models shape the priority list?

It is not settled. What the evidence supports, and what most experienced clinicians believe, is that the role becomes more central, not less, but central differently. The offloading of menial, time-consuming tasks is real and permanent. Quick access to the entire human archive is real. What remains is synthesis, judgment, relationship, and accountability, the irreducible core this book has been arguing for. Whatever else the coming decade takes off the clinician's desk, it does not remove the Diagnostician.

This book has shown that technology changes decision-making and care delivery for better and worse. The most direct statement I can make is this: **physicians and other clinicians who over-rely on AI for decision-making, offload tasks that they should be watching, and yield judgment to a black box instead of a professional method, will find it hard to justify their market worth.** If we graduate clinicians whose professional ethics and cognitive gifts are allowed to decline, offering nothing more than the tools around them, some specialties face an extinction event. That needs attention at the academic and policy level now.

Designing clinical roles, training, and team structures that protect that core while taking full advantage of the tools is an informatics design problem, and it belongs to a team that understands both the clinical work and the machinery. (See Figure 16.1.)

THE PROGNOSIS · THE LONG COURSE

Healthcare's Prognosis

Five conditions arriving on a schedule nobody controls.

1	**Payment Squeeze**	Reimbursement now rewards outcomes and risk data most organizations still don't capture.
2	**AI Sprawl**	Predictive, generative, and agentic tools are multiplying faster than anyone is governing them.
3	**Workforce Squeeze**	The clinical labor that once absorbed bad workflow is gone, and the shortage is still widening.
4	**Record Flood**	Patient records are finally moving between organizations, and almost nobody has built the workflow to use them.
5	**Role Erosion**	As systems absorb documentation and coordination, the clinician's irreducible core goes undefined and unprotected.

THE RATE-LIMITING STEP

Not money. Not technology. People.

ABOUT

50

Graduates a Year

The practice pathway closed after 2025. No other door remains.

FIGURE 16.1

Five conditions, one bottleneck. All five are answered by the same person.

THE PROGNOSIS FOR CLINICAL INFORMATICS

Every condition above is met, or failed, by the same capability: the informatics function. The supply of that capability is the number that determines all the others.

From this year forward there is one route: 24 months of an accredited fellowship (ABPM, 2025). So far the fellowships cannot fill.

For the 2024 to 2025 academic year, the Accreditation Council for Graduate Medical Education reports around 50 graduates in the last tracked year (ACGME, 2025). That is the annual national production of the only credential that now qualifies a physician to hold this subspecialty, for a country with more than 6,000 hospitals.

Only 60% of this country's training capacity is filled. The seats exist. The accreditation exists. The faculty exists. The demand doesn't.

The Journal of the American Medical Informatics Association (JAMIA), the main journal of the discipline, published a survey of clinical informatics fellows in 2023, and its denominator is the useful part: the authors identified 394 alumni and current fellows across the graduating classes of 2016 through 2024 (Kim et al., 2023), the whole fellowship-trained cohort, 9 graduating classes. Under 400 people. At 50 a year, placing one fellowship-trained physician in each American hospital would take longer than a century, and that is assuming that there are no retirements, departures for industry or academics, and no changes of mind.

When the practice pathway first opened in 2013, 445 physicians passed the examination in that single sitting (Lehmann et al., 2018a; Detmer and Shortliffe, 2014). One day in 2013 produced more certified clinical informaticists than the fellowship system has produced in its first nine years.

Chapter 8 found that 7 in 10 organizations already claim this role to be filled on their org chart. Those two facts, the training volume and the workforce prevalence, cannot both describe the same people. About 70% of organizations hold the title, which is an estimated 4,200 positions, while the country has produced fewer than 400 fellowship-trained physicians in nine years.

The workforce issue becomes clearer. There were never enough trained people for the prevalence numbers to mean anything. The Name spread because it is free. The Form and Function did not, because they are not: it takes science, technology, engineering, and mathematics (STEM), medical, and informatics training, experience, graduated responsibility, and decision-making authority.

That figure reframes the problem. The constraint is not a shortage of programs. The key constraint is that a physician finishing residency will not take a two-year salary cut to train for a subspecialty most department chairs have never heard of and most hospital executives cannot define. Especially with no documented compensation premium at the end of it, in a field whose own value has never been published (Detmer, Munger, and Lehmann, 2024; Lehmann et al., 2018b).

None of American healthcare's illness improves without a sound informatics function. Thousands of hospitals are running thin on margin. Risk contracts are requiring clinical capabilities that have never been built. A growing AI footprint that will exceed the entire volume of today's alert libraries is coming. Healthcare has a clinical workforce that can no longer absorb inefficiency.

And, the new technology keeps arriving. The rate-limiting step in American clinical informatics is not money, not technology, not executive willingness. It is the available workforce, and few organizations are even screening for it.

CLINICAL INFORMATICS: THE SECOND DOOR

The need is proven. Here is the intervention.

A second door is what a profession builds when the front door is preferred and cannot pass enough people through it in the time available.

It is not a lowered standard. It is a temporary response to a real shortage that keeps the front door open the entire time.

Medicine has done this repeatedly. It is how the country produced the hospitalist, how nurse practitioners scaled into primary care shortage areas, how rural coverage has been sustained for decades. The argument was identical every time; the formal pipeline was correct, and slow, and the need could not wait, so alternate training ran alongside it until it caught up.

Executive practice also does this constantly. Interim chief financial officers. Fractional general counsel. Shared chief information security officers. Contracted operating executives brought in for a turnaround. No board treats any of it as unusual. It is ordinary action, deployed when a seat cannot sit empty and cannot yet be filled permanently.

This is not a novel proposal. It is the standard response to a workforce gap, applied to a seat few have thought to apply it to.

It is, I think, the best available answer to what this chapter has diagnosed. Too few formally trained graduates can meet the looming demand on a timeline that saves the patient. The second door is not a supplement to the solution, for most of the next decade it is the solution, and the fellowship is what makes it permanent afterward.

Five Treatments for the Shortage

A physician waiting for the executive seat, or already sitting in it, can be developed in place by a Diagnostician who understands the function, has seen the illness, and has treated it before. It is not certification, and is not presented as one. What it produces is competence in the informatics function, a different thing than a credential, and the thing organizations need most now.

No organization that recognizes the need today can wait 2 years for the academy to produce the person, and most will not pay to send a sitting physician away for that long. This is not an argument for lowering the standard, only the observation that one narrow door admits too few people.

Chapter 10 conducted a morbidity and mortality conference over the broken candidate pipeline. This is the part where we have to say what will be done differently.

Development takes time that a struggling organization may not have. Five arrangements can hold the informatics Function in the interval or become the long-term answer; they overlap and can be combined.

The interim seat. The organization contracts a skilled Diagnostician, named temporarily as the accountable executive, while it builds or recruits for the permanent role. The initial work on critical local needs can begin; the initial scaffolding for the informatics Function can be laid; or even the test drive of the candidate for the permanent role can be initiated.

Unfortunately, an organization that cannot recruit this Function rarely goes without it; they give the title to some candidate with some administrative time and then call the problem solved. A vacancy is at least named honestly. A wrong appointment is sometimes just a vacancy in disguise.

Remote coverage. The organization contracts a trained Diagnostician to hold the executive role, warm the seat, mentor the future, or attack an identified issue, from wherever they are, with their authority made clear. This is the arrangement with the strongest evidence behind it.

The shared seat. Several small organizations fund one trained executive between them. Rural America has run clinical specialty coverage this way for four decades. A four-hospital consortium can jointly afford a person that none of them can afford alone, and each of them currently has the title on their organization's chart.

Supervised internal promotion. The organization promotes its own physician into the seat immediately and contracts an experienced Diagnostician to supervise the work rather than to do it. The seat is filled from day one and the competence is built while it is occupied. Many organizations should be considering this and few are.

The targeted engagement. Not a seat at all. It is one or more defined engagements targeted at defined local problems, with a written finding at the end. They can be hired for the assessment, the diagnosis, the treatment plan, or the cure. Targets include the alert library. The AI inventory. The denial pattern. The strategic plan. This is the smallest organizational commitment on the list, and frequently the one that shows a board what the informatics Function can accomplish.

COVERAGE AT A DISTANCE

The shifting U.S. workforce has been forcing some fundamental changes in staffing models. Shortages are being increasingly filled with resources who live remotely from the organization. Remote coverage sounds like a compromise to people whose experience has been only in-person. For some roles, it is not, and the evidence is growing.

The first is this book's own, and is a clinical example. Chapter 14 examined Kalvelage and colleagues, who pooled 477,637 patients across 20 studies of remote critical care. The variable that mattered most was not the location of the physician, it was the installation of the role into the organization. Where the remote intensivist team held decision-making authority, mortality fell. Where the same team could only observe and advise, not decide, the benefit vanished.

The variable was the recognized authority to act. American medicine settled that question in critical care and emergency departments long ago for the sickest patients in the building, and few have yet to carry the finding one floor up.

The second comes from outside healthcare, and it is the largest randomized trial of hybrid work ever conducted. Bloom and colleagues randomized 1,612 professional employees to full-time office work or to 3 days in the office with 2 days remote and followed them for 6 months. Quit rates in the hybrid group fell by roughly a third, from 7.2% to 4.8%. Performance reviews, promotion rates, and measured output showed no difference at all (Bloom et al., 2024).

Retention improved. Performance did not suffer. That is a randomized trial published in Nature, and it describes the arrangement being proposed here.

Almost all informatics work transmits well from a distance: committee and governance meetings, particularly ones where half the participants are already remote; design and build review; data and analysis; policy, charters, and written findings; vendor management; teaching, mentoring, and coaching, which transmit unusually well because they are conversation and review.

Some things do not transmit as well, often paralleling the limitations telehealth has experienced. Rounding, the physical examination described in Chapter 15, cannot be done as well through a screen. Relationship-building at the outset, before the organization has a reason to trust the person, benefits from in-person time. Reading a room during a hard decision is harder. Being visibly present when something is going wrong is not always possible.

Those are real concerns; they are also addressable and schedulable. What works well is a mixed route. Schedule defined intervals on site for the work that requires a body in the room; everything else can be delivered at a distance.

One more benefit belongs to the entire healthcare market. Relocation is the most common reason a qualified candidate declines any seat. Demanding a move degrades the candidate pool; a remote model enlarges the search to a national pool, and that helps all involved.

WHAT WOULD CHANGE THE PROGNOSIS

A prognosis names what is modifiable. Some of this belongs to the organization and some of it is outside of its control.

What your organization can change, starting now

The Diagnostician seat filled and empowered. Not the Name, the six success conditions from Chapter 8: education and preparation, funding, positioning, empowering, staffing, and employing.

One of the five Diagnostician arrangements above, contracted to attack an issue this quarter. Interim, remote, shared, supervised, or a targeted engagement. Do not wait a year on a search against a small pool.

Governance and workforce sized to the load of work you are about to have, rather than the one you have now. What is arriving is significantly larger than what is here, and staffing has a multi-year lead time.

Workflow reconciled with the changing environment. The new data, transforming technology, and field changes are coming whether you have designed for them or not.

The executive team taught the clinical method. Chapter 15 argued this is cognitive and trainable, and it is the cheapest item on this list by an order of magnitude.

What only the nation can change

Four things sit above any organization's reach. Chapter 17 addresses them specifically.

National executive education in the Diagnostician role, the function, and the return. Every executive team in America can name what a cardiologist does and roughly what one is worth. Almost none can do it for the Diagnostician, and no institution currently teaches them.

A national commitment to producing the Diagnostician and informatics function. Not only accrediting more programs. Funding the seats that already exist, publishing the career path, and supporting the second door with standards attached.

Deliberate use of the second door while the first one is widened. Interim and remote coverage treated as legitimate national capacity rather than as local improvisation, and remote arrangements permitted wherever the work allows, because relocation is what shrinks the pool.

National recognition of how sick this patient is. How much it costs, how much it harms, how many people it drives out of the profession, and why continuing without a change is not the neutral option it is being treated as.

That last one is the argument of this whole book. The Diagnostician and the clinical method are the answer to it, and Chapter 8 priced what the answer is worth.

Every item is modifiable. None requires another technology purchase. All of them require a Diagnostician with the training to know what to do and the standing to do it.

BEDSIDE QUESTIONS: THE PROGNOSIS OF YOUR LONG COURSE

1. Have you characterized the value-based capabilities you will need at your next major contract renewal, and budgeted for them, or are you planning to build them during the contract?
2. Is your governance sized to the number of AI models you will be running in 3 years, or the number you are running now?
3. Is your workforce strategy reconciled with your

technology strategy at the executive level, in the same room, with the same people?

4. Are your interoperability and population health capabilities designed for the cross-organizational future that has already begun, or for the within-organizational past?
5. Who is your informatics succession plan? If the person in that seat left in 90 days, what would you do? Now check that answer against a national graduating class of around 50.
6. Are you developing the physician already in the seat, or waiting to recruit one who is already finished? If waiting, how long has the search been open, and what is your plan for the year after it fails?
7. If your search has already failed once, which of the five arrangements did you evaluate before concluding you could not have this function? If the answer is none of them, the conclusion was premature.
8. Is your residency and training partnership designed for the physician role that is arriving, or the one that existed when the partnership was written?
9. When your board asks what the next 5 years require, can your local team give a coherent answer without a vendor in the room?

If you cannot answer these, there is no long-term operating plan for the informatics function, the role, the workforce, and the tuned technology.

THE PROGNOSIS HAS BEEN RENDERED. FOUR ONCOMING healthcare changes named, arriving on an uncontrolled schedule, all of them can be met squarely by the same capability. A rate-limiting step has been identified, which is not money or technology, but people. And, five arrangements that can hold the Function while you prepare for permanence.

The patient is more complicated than the local environment can handle. Let us finish with a referral to the highest level.

CHAPTER 17

REFERRAL TO A HIGHER LEVEL OF CARE

THE NATIONAL INFRASTRUCTURE FOR THE EHR

Some conditions cannot be treated where they present. The physician who recognizes one has a specific obligation, and it is not to keep trying. It is to write the referral, state the findings plainly, and name what is being asked for.

The head injury alone is not what worries the physician on call at the rural hospital. It is the availability of the CT; there is no neurosurgeon within eighty miles, no operating room staffed for a cranial case, and no version of this evening where waiting to see what happens is the responsible choice.

So, the physician does everything that can be done here. The airway is secured. The blood pressure is brought down. The blood thinner the patient takes is reversed. A bed is found ninety miles away, at a hospital with a neurosurgeon and a helicopter that can be there in eleven minutes, and the physician calls ahead personally.

The note that travels with the patient is not a summary of the night, it is an instrument. It states plainly what was tried and what it did. And, it states just as plainly what is still needed; neurosurgical evaluation on arrival, an intensive care bed, likely evacuation of the bleed within hours. Recognizing the edge of your own capability is not a failure; it is visible competence.

THE REFERRAL

A referral is both an admission and a demand all at once. It is the admission that the problem exceeds the local capability, and the demand is for a clinician with more capability to take over care. A good referral states what was found, what was tried, and what is being asked for; it is not simply a request to be rid of a problem.

Sixteen chapters have described a disease, named a diagnosis, priced the treatment, and prescribed it, nearly all of it treatable by people already on an organization's own payroll. What remains is not local disease. No health system can fix the environment from inside its own walls.

This chapter is that referral letter, addressed to the people who hold what an individual health system does not: the ones who write the rules and the ones who pay the bills, who, unsurprisingly, turn out to be the same people.

THE FOUNDER'S NOTE LEFT IN THE CHART

In July 1959, Ledley and Lusted published the paper generally credited as founding informatics, "Reasoning Foundations of Medical Diagnosis" (Ledley and Lusted, 1959). The field was born 67 years ago as a question of how physicians reason, not how medical records are stored or claims transmitted. The clinical method this book has argued descends from that original charter.

9 years later, Lawrence Weed argued in the *New England Journal of Medicine* that the medical record is an instrument of reasoning, not a business "container." And, that a record built for any other purpose would corrupt the thinking of every user who touched it (Weed, 1968). He published that warning in 1968, 41 years before HITECH.

The founders kept asking important questions.

Warner Slack argued in *The Lancet* that a patient has the right to decide about their own care rather than be managed by others, at a time when hospitals routinely left them out of that decision (Slack, 1977). And, 49 years later that question is only now approaching settlement.

In 1991, an *Institute of Medicine (IOM)* committee called the computer-based patient record an essential technology, organized around the patient rather than the institution (IOM, 1991). We are still approaching that 35 years later.

In 1986, the *National Library of Medicine* began building a common clinical vocabulary, the Unified Medical Language System (UMLS), so that a diagnosis would mean the same thing in different systems (Lindberg et al., 1993). A language that 40 years later is still inconsistent.

In 2005, Ed Hammond published in *Health Affairs*, the journal American health policy actually reads, exactly why data standards were required and what stood in their way (Hammond, 2005). That paper ran 4 years before HITECH. Policy proceeded without the standards and the country is still paying for it 21 years later.

Underneath all of it sit the same three instructions: understand the work before you automate it, design for the clinician's cognition rather than the institution's convenience, and train physicians in the informatics discipline formally, because it sits at a junction of the new medical reality. These were clinicians and scientists, not administrators chasing efficiency.

They believed that if you could formalize how a good physician thinks, you could offer that to every patient in the country. It is one of the more generous ambitions in American medicine and predates HITECH, some by 30 years.

None of it was done first.

THE DIAGNOSTICIANS THAT WERE NEVER CALLED

The National Library of Medicine began funding clinical informatics training in 1972 (National Library of Medicine, 2018). By the 1990s it had produced a small, identifiable population of physicians trained in exactly this discipline.

Small is the operative word. There was no movement. But they existed and were findable, and the federal government had already paid for them.

When the United States decided in 2009 to digitize American medicine, the founders were in their professional prime, and their arguments were sitting in print in the journals policy-makers actually read. A staffer with a library card could have found all of it.

Physicians sat on the program's advisory committees. They never referred the case to an expert.

The specialty was not asked to evaluate the case before the treatment was chosen. A recent consultation was available. It was federally funded, spanned decades, multidisciplinary, and never requested. That is the part of this history that is most relevant, and the reason this chapter exists.

WHAT WAS ORDERED INSTEAD

In 2009, Congress appropriated roughly $32 billion to put an electronic health record (EHR) in every American hospital and physician office (Blumenthal, 2010; Buntin et al., 2010).

No history was taken. The actual work of patient care went unobserved, and the instruments clinicians would be required to use on their patients were designed anyway. No differential diagnosis of the industry illness was generated. The assigned diagnosis was named physician technophobia and low adoption, and adoption became the only thing measured. Inside an industry on the nation's technologic cutting edge.

The treatment was already prescribed before the diagnosis, the intervention chosen first and the criteria written afterward, built to measure what the primitive tools could collect rather than what they should collect. The endpoint was a process measure. "*Meaningful use*" counted as a checked box; not a healthier patient, not a clinician able to think, not data which meant anything. And, when they finally checked whether patients were better off, the effect on quality was small and uneven (Appari et al., 2013).

Chapter 2 showed what happens to any measure that becomes a funded target, it gets chased whether right or wrong. This was that, at national scale, with tens of billions behind it.

There were no discontinuation criteria, response that would cause the program to stop, revise, or reconsider. The clinical outcome had no accountable expert. The program had administrators, contractors, certification bodies, and vendors, and not one Diagnostician reading the whole picture.

THE PAYER WHO WROTE THE PRESCRIPTION

Now the structural problem underneath all of it. The Centers for Medicare and Medicaid Services (CMS, 2025c) insures roughly 39% of American health spending directly, and its documentation and coding rules get adopted by the commercial payers who cover the rest. CMS writes the rules for the whole country, whether it is paying the bill or not, and it does so from three conflicting seats at once. It specifies what the EHR must record to get paid. It pays providers based upon that document. Then, it audits that document when it disagrees. Specifier, payer, and auditor with no independent party anywhere in the loop.

In any other regulated industry that arrangement gets challenged. In healthcare it was accepted, and it remains the arrangement today.

Weed predicted the result in 1968. A record built for any purpose other than clinical reasoning will corrupt the reasoning of those who use it. He did not know who would eventually build it or why. He did know what would happen once they did.

The timing made it even worse. HITECH passed in February 2009, during one of the worst financial crises since the Depression, when hospitals were capital-starved and margin-starved and in no position to decline the money or take the time to think it through. The permanent national recovery audit program launched the following January, expanding documentation requirements and enforcement (CMS, 2025a). A deadline-driven mandate, arriving when organizations could least afford to be careful, from an agency that would later audit the output, with penalties reserved for any organization slow enough to be careful.

SEVENTEEN YEARS OF FOLLOW-UP

Every treatment earns a follow-up visit. This is the chart.

One vendor now holds 43.7% of American acute-care hospitals and 56.9% of all hospital beds (KLAS Research, 2026b). That consolidation was never the stated plan. Within 3 years of the incentive's start, several hundred products carried EHR certification. Narrow specialist tools, a handful of enterprise systems, that on paper looked like the marketplace the policy intended. In practice, just 6 vendors already accounted for 3 in 4 of all hospital attestations (Wanderer, Mishra, and Ehrenfeld, 2014).

By the government's own antitrust yardstick the hospital EHR market soon crossed the threshold into "highly concentrated," and it never looked back (Holmgren and Apathy, 2023). Today 2 platforms hold close to 66% of the acute-care market between them, and 3 platforms combined serve more than 80% American hospitals (KLAS Research, 2026b; ONC, 2026).

Competition was outbid with public money and never recovered. The price keeps rising, and the buyer has no alternative.

The data stayed poor, because the record was built for billing and attestation rather than clinical reasoning, and that is what trains the AI models arriving next.

The clinicians promised relief received the largest documentation burden in the history of the profession, governed by technology they were never trained to read. Seventeen years, tens of billions of dollars, and the patient is poorer, more exhausted, accumulating worse data, inside a more concentrated market, with no accountability for the clinical result.

In any other domain of medicine that record triggers a case review. In this one it produced silence.

THE UNTAKEN HISTORY

During all of this, the clinical informatics function has never been accessed.

In 2016, an AMIA task force published one of the most serious accounts available of the Diagnostician's role, defining the knowledge and skill a *chief clinical informatics officer* needs (Kannry et al., 2016). Inside sits the issue the field has never reckoned with: the role remains heterogeneous in title, responsibility, and scope, based upon which local organization invents it.

The discipline's own leadership said, in print, that no consensus exists on the job. That was 10 years ago. Since then the field has written an official curriculum, built a fellowship, defined competencies, and administered an examination. It has never followed up. There is no national census of where these people sit, who they report to, how much protected time they hold, or whether they can halt an operational deployment. The figures of organizations with an identified CMIO come from a workforce survey asking only whether a senior clinical information leader is employed (HIMSS, 2019).

Employed, not empowered, not resourced, not positioned. I know a CMIO who sits in the academic wing with no operational authority over the EHR, another with a title and four hours per week, and another who is a C-suite executive who can halt a go-live. All are counted in the same 7 in 10. The central claim of this book, that the Name is common and the Form and Function are rare, rests on an absence of any discernible measurement. No one has looked.

THE REFERRAL FOR A CONSULT

First, *require a diagnosis* before writing a national prescription. State the problem in advance, what other options were considered and ruled out, what outcome will be measured, what result would change course, and who is clinically accountable. That standard already applies to a second-year medical resident with a patient having chest pain. It has never applied to a national health technology program, and we have run several.

Make every quality measure computable. A measure that requires a human abstractor was designed badly, and the cost lands on resource time that the enterprise is paying for.

Publish the denial arithmetic. Every payer, every year; prior authorizations issued, denials, and the share ultimately overturned. Publication alone will move behavior.

Take the census, then run the study. Survey what CMIOs actually do, publish it, stratify organizations by how much of the Function is genuinely present, and measure the outcomes Chapter 8 priced. If the association is absent, I am wrong, and the field will have learned something. Seventeen years of no findings about this role are not evidence the role fails. They are evidence that the dose was never measured.

Consult the specialty before the next program. This is the smallest request in this chapter.

These problems went unsolved this long for a simple reason. Research and reform flow toward what their sponsors understand, hospitals, physicians, medicine, surgery, the business of care, the patient in the room.

Informatics is the hidden discipline underneath all of it, the science of making the data trustworthy enough for everything else to work. The failures stay invisible until something breaks, so it stays chronically neglected.

Three of its foundations remain unfinished.

WHOSE RECORD IS IT?

At the bedside, we confirm the identification of a patient with two identifiers before we act. Scaled to the country, that habit collapses. The United States has no reliable way to know that the John Smith treated in Miami is the John Smith admitted in Denver. Records match correctly as little as 80% of the time inside a single institution, and as little as 50% of the time between institutions (ONC, 2014; Pew, 2018). ECRI reviewed over 7,000 identification error events and found that roughly 9% caused significant harm, including 2 deaths (ECRI, 2016). A mislabeled specimen is the laboratory's cardinal sin. A misidentified patient is the same error, one level up.

HIPAA called for a unique health identifier for every American in 1996 (HIPAA, 1996). It was never built; an appropriations rider has forbidden the federal government from spending a dollar on one every year since 1999 on privacy grounds. The private sector has shown what is possible anyway. Intermountain Healthcare, working with its partners in the Care Connectivity Consortium, drove its own patient chart match rates from the single digits to better than 95% through disciplined governance alone, a result the Sequoia Project later published as a case study for the cross-organizational framework it built with that consortium (Sequoia Project, 2018). Identity is the keystone. Without it, every inference downstream inherits a margin of doubt.

A COMMON LANGUAGE

Medicine speaks in many language pieces.

Diagnoses live in ICD-10-CM, stewarded by the Centers for Disease Control and Prevention (CDC).
Inpatient procedures live in ICD-10-PCS, stewarded by the Centers for Medicare & Medicaid Services (CMS).
Physician services live in CPT, owned by the American Medical Association (AMA).
Clinical concepts live in SNOMED CT, distributed by the National Library of Medicine (NLM).
Lab results live in LOINC, curated by Regenstrief Institute.
Imaging lives in DICOM, managed by National Electrical Manufacturers Association (NEMA).

The National Library of Medicine stitches these vocabularies back together after the fact.

Each is excellent. The seams are the problem. They have different owners, different release cycles, and different purposes: billing in one dialect and clinical care in another. Translating faithfully between them is constant, costly, and imperfect work.

The rest of the world moved to the 11th revision of the International Classification of Diseases in 2022 (WHO, 2022). The United States still runs the 10th. Countries that cannot agree on the basic words for a disease will struggle to compute on them.

THE RECORD THAT FOLLOWS THE PATIENT

A diagnosis is a trajectory over time, not a snapshot, and an EHR that cannot follow the patient across settings cannot support the reasoning it exists for. Here the country has made real progress. The 21st Century Cures Act made blocking health information unlawful and ordered it exposed through standard interfaces. The implementing rule required certified systems to expose data through a common application programming interface (API) built on Health Level Seven (HL7) FHIR (ONC, 2016; ONC, 2020). The Trusted Exchange Framework, live since the end of 2023, has already carried more than a billion records across its network of health information networks (HHS, 2026).

Moving documents is not the same thing as assembling a life. What exists today is a record exchange between completely isolated charts. What medicine needs is one coherent, person-centered record that follows a patient across a lifetime and many institutions, the argument W. Ed Hammond himself is now making for what should replace the institution-centered record he helped build (Hammond, Bent, and West, 2022). When the people who helped build the first electronic record are calling for its successor, the rest of us should listen. (See Figure 17.1.)

THE NATIONAL TIER · BEYOND THE LOCAL ORGANIZATION

The Three Unfinished Foundations

What no single hospital can fix, and the leadership it requires.

1 Patient Identity

THE GAP No reliable way to match one patient's records across organizations.

TODAY As low as a 50% match rate when institutions exchange data.

2 A Common Language

THE GAP Clinical and billing vocabularies were each built well, but governed apart.

TODAY The rest of the world moved to ICD-11. The U.S. still runs ICD-10.

3 The Longitudinal Record

THE GAP Moving documents between organizations is not the same as assembling a life.

TODAY Records now move at national scale. Nobody has assembled them into one chart.

THE ATTENDING FOR THE SYSTEM

These are problems of will, not insight. They yield only to the executive-physician, one tier up – clinical reality, technical possibility, and executive judgment, held in a single mind, accountable for the whole.

FIGURE 17.1

The three foundations no single organization can complete, and the executive-physician who must.

THE DIAGNOSTICIAN FOR THE SYSTEM

None of these three problems, identity, language, or continuity, is a problem lacking insight. We know what is wrong and largely know what to do. They are problems of will and coordination, sustained across vendors, institutions, payers, and decades, and they will yield to the same kind of leadership this book has argued for all along, exercised one level up. The nation needs its own Diagnosticians seated where the rules get written, in the federal technology offices, the standards bodies, the payer and policy tables; people who can hold clinical reality, technical possibility, and executive judgment in a single mind, and who are accountable for the whole of it.

This is the hardest part of the work to fund, because it is invisible to most until it fails, and it is also the substrate everything else depends on. A *learning health system* cannot learn from data it cannot trust, cannot read, and cannot assemble (IOM, 2012). Get identity, language, and continuity right, and the data hold at the scale of a country. Get them wrong, and the most sophisticated hospital in the land is just guessing.

BEDSIDE QUESTIONS: QUESTIONS FOR THE FIELD

1. Can your organization state, with evidence, its patient-matching accuracy; both within its walls and when it exchanges data with others?
2. Who, by name, is accountable for terminology governance, and how are the mappings between billing and clinical vocabularies maintained?
3. When a record arrives from another institution through TEFCA or a regional network, does it become part of a

coherent longitudinal record, or merely another document in a pile?

4. Which national bodies; standards organizations, and your specialty societies, set the rules you must live by, and are you represented where those rules are made?
5. If a unique patient identifier were permitted tomorrow, is your data governance mature enough to use it well?

You likely join the vast majority of the industry in having a set of blank answers. They are not issues you can fix. They are only issues that you can advocate for.

WE HAVE RUN THE WHOLE WORKUP ON THE PATIENT, PROVIDING a prognosis and outlook. We have done the work at as high a level as can be achieved in our local community. And, we have shown that there are diseases nationally, outside of our purview, that need to be referred to a Diagnostician specialist at that level. The Conclusion is the discharge plan. The patient is ready to go home.

CONCLUSION

THE DISCHARGE PLAN

A patient who has been carefully diagnosed, properly treated, and adequately monitored does not simply walk out of the hospital when the acute episode ends. The physician sits down with them and goes through what happens next.

Mrs. Smith, a 42-year-old woman with type 2 diabetes diagnosed 3 years ago, comes to clinic for her regular followup.

Before she sits down, the physician already knows three things she does not. Her hemoglobin A1c came back at 7.4% last week, drifting upward. Her glucose monitor data, pulled into the chart automatically, show a pattern of post-dinner highs she has not noticed. And, her kidney function, ordered annually as a structured element inside the order set rather than as just a memory, has just crossed into the early warning range.

The physician does not go looking for any of it. The record surfaces all three before the visit opens, on a layout the informatics executive and the diabetes team built together.

Decision support, which used to fire a dozen irrelevant alerts during a diabetes visit, fires three. Consider a GLP-1 agonist for weight loss. Discuss the post-dinner pattern with the patient. Review the kidney function with her and what she needs to do in response.

The physician reads all three.

The note is captured by an ambient scribe configured earlier that year to draft the way this physician actually thinks, reviewed, corrected, and signed in 8 minutes instead of 30.

The patient leaves with a clearer plan than she has had in years.

She never saw the technology and never thought about it. It worked because the organization had been doing the difficult work for several years before she walked in.

That visit is the deliverable. Not an aspiration and not a vision statement. The difference between that visit and the one that is happening right now in most American organizations is not technology and it is not talent. It is a Diagnostician, governance, and accountability.

If you have read this far, you have completed the diagnostic episode for your own organization. Chief complaint, focused history, differential, targeted workup, treatment plan, follow-up cadence, long course. What remains is the discharge plan, and a discharge plan has four parts. What to do now. What to do over time. What to watch for. And, when to call back.

A WORD ABOUT TIME

The horizons below are goalposts, not guarantees. Where an organization lands on the maturity scale depends on what has already been done or left undone.

Most organizations arrive here already behind. Installing the EHR software is hard work, but it is less than a third of the effort of a modern go-live. The vendor will deliver a working system on the agreed date with near-perfect reliability. The contract makes it clear that the client owns every decision, every unattended meeting, and every misunderstood setting choice. They will engage you as though you know what to do and your governance was already strong, and they will hit their date whether it is or not. A hospital paying over $200 million for that software gets much less deference than it used to. Every party involved, except the organization itself, recognizes its permanent dependency on outside help to install and maintain the system. The initial investment will seem small within a few years of maintenance costs. Even more, the exit cost assures that it cannot go anywhere else.

American healthcare has earned every bit of that treatment because it has been mistakenly hoping that a platform, not hard work, would compensate for years of neglect. The organization follows the rules and the EHR goes live exactly on time.

Then, the vendor and consultants go home.

The future plan differs everywhere by its starting point. How much of the work this book discusses has already been done?

The rare organization starting from scratch with a new EHR, who brings in the right Diagnostician, can save itself years. Most organizations installed their EHR without a Diagnostician and face years of rebuilding what they already paid for. A few organizations were tended well by a Diagnostician from the start and need only to add the workforce and strategy. All three paths take years, because budgets and culture both move slowly, and an organization that does not recognize its own predicament will not ask for the help it needs.

Every organization performing well today started somewhere, 3, 5, or 7 years ago. Here is where you can start.

DISCHARGE ORDERS: THIS QUARTER

First 90 days. Beginning is an act of leadership and not a technical one.

Score yourself in writing: The six conditions for Diagnostician and informatics success from Chapter 8: education and preparation, funding, positioning, empowering, staffing, employing.

Compute your financial exposure, your physician count times the figure in Chapter 8, and put it on the same page as what you currently spend on the informatics function. That is one afternoon, and it is the entire business case.

Name the informatics Function as a strategic priority and fund it, not the title: a positioned Diagnostician, initial staff, real participation from the rest of the C-suite, and enough communication that the organization knows a new priority has arrived. This costs $4 on every $100 you spent on the EHR software.

Begin the search for a Diagnostician immediately. Then, choose from five interim-coverage arrangements while you search: an interim seat, remote or fractional coverage, a shared seat with a neighboring organization, a supervised internal promotion, or a targeted engagement against one defined problem.

Identify the physicians you are going to develop. There is almost certainly one with the interest. It is time to speak to them.

None of this completes the recovery. It makes the recovery possible.

DISCHARGE ORDERS: THE FIRST YEAR

A search for any executive takes six months to a year, longer if there is a shortage and the house is not in order. With a capable interim in the seat, the first year builds the foundation: infrastructure, workforce, and the 3-to-5-year plan.

All of it depends on the local culture. Is the chief executive actually in charge? Is the board more than a placeholder? Is there real competence in the existing informatics structure and leadership? Is this a priority or a whim? Where culture is not ready, the work costs more and takes longer. That is a reason to start now, not to wait.

DISCHARGE ORDERS: NEXT FIVE YEARS

By five years, the measure is how completely the Function is woven into the organization's culture: a place on the executive scorecard with real metrics and real accountability, access to the data and the AI models, authority to act on what it finds, a C-suite that understands the stakes, and physician leadership that is accountable rather than merely consulted.

Everything in this book fits inside that window when the commitment is genuine. There are organizations that have already done the work, they do exist, and they will usually welcome your visit. What tends to surprise people the most is how little of that five years is spent on acquiring new technology; it is best used to develop the disciplines described in this book.

MAINTENANCE

The work does not end, it transforms the organization's character. A mature informatics Function adapts continuously, because medicine, disease, and technology will keep moving. The work and mentoring help develop the next Diagnosticians from the inside, creating a pipeline. At that point of maturity, optimization is not a new project or a staffed campaign. It is the organization functioning well, as its culture.

WHEN TO CALL BACK

Every discharge plan identifies the signs and symptoms that signal recovery is failing.

The seat goes unfilled and progress has stopped.

The seat is filled and has not stopped a decision or retired a clinical workflow.

The role's protected time gets borrowed by other clinical, IT, or revenue cycle initiatives; the first time it serves another leader's agenda, the resources will be borrowed and distracted permanently.

Governance meets and decides nothing. A committee that only meets has no function.

The Diagnostician's interventions have no discontinuation criteria. Every plan can be wrong. Measure, and adjust course.

Local informatics development has not occurred. No successor is named. This becomes a single point of failure waiting to leave.

Any one of these means that the recovery has stalled, and it has yet to be recognized.

THE FOLLOW-UP FOR THE READER

A discharge plan says who does what. Seven groups can act on this book, one act each, available to whoever has read this far.

The Patient: ask who is accountable for whether the technology used helps the person treating you. You are entitled to an answer.

The Clinician: stop absorbing EHR malpractice quietly. Every workaround that keeps your clinic moving also keeps the defects invisible to whoever could fix them. Report the issues before they become an incident.

The Executive: take one of your key dashboard metrics and audit it from the beginning of the clinical workflow to the end.

The Board Member: ask for every operational number on one page, financial exposure and current spend. Pay special attention to the Diagnostician and clinical informatics upside.

The Regulator: require a diagnosis before a national regulatory prescription, and consult a Diagnostician to avoid negative downstream consequences.

The Payer: every requirement you write is executed by a clinician inside a system that was designed and configured by others. Ask what it costs to produce before you require it.

The Legislator: people are the rate-limiting step in most health-care issues, the rare problem money actually solves. Fund the creation of Diagnosticians and permit the unique patient identifier the government has been forbidden to build since 1999.

WHAT THIS FUNCTION IS FOR

A book about this role, at the end, must be a book about leadership. The person in this seat is a Diagnostician whose specialty is the junction of clinical work, technology, workflow, and data. They are accountable for synthesis with the rest of the executive team and clinical enterprise. They hold the whole picture, write the unifying plan, and answer for the coherence of the clinical technology environment the same way the physician running a complex case answers for the patient in the bed.

This is not a luxury. It is the missing executive in most American healthcare organizations, and moving from an organization that lacks the role to an organization that has it is among the most consequential leadership investments available in the next five years, at a cost of $0.14 (new money) to every $100 the absence already costs.

That is the clinical argument. What follows is the personal one.

A FINAL WORD FROM THE DIAGNOSTICIAN

I HAVE SPENT THIS ENTIRE BOOK BEING CAREFUL.

Careful with the numbers, careful with the sources, careful to concede what could be conceded and to state where I might be wrong. That care was intentional. An argument this large, formed by a writer with interest in it being true, survives on its arithmetic and objectivity.

The arithmetic is behind us now. What follows is the part I have kept out of the seventeen chapters.

I trained for this profession before there was a well-known name for the role. I was motivated by the Function and its importance.

In the 1990s I went into medical informatics because I could see, in a way I could not yet explain, that the tools were going to decide what kind of medicine we would be allowed to practice. A small number of us saw it. We were not psychic. We were simply looking at the right thing at a time when almost no attention was being paid to it.

Then I spent a career watching the things we were afraid of happen anyway.

I want to frame this better. My career has been good. I have worked with excellent people. I have had the seat, wrestled for the authority, and had to beg for the budget, which is more than most of the people this book is written about will ever be given.

What I have not had, and what none of us has had, is the country's attention.

This is what that has actually looked like, in the ordinary week.

It looks like watching a physician who was excellent at their job become slower, then unhappy, then absent, and knowing precisely which configuration decisions did it. All were made in a room I was not invited into.

It looks like reading an incident report, recognizing the mechanism immediately, and understanding that the same mechanism is live in eleven other places in the same building.

It looks like presenting a finding to a group of intelligent, decent executives who nod, agree, thank you sincerely, and then ignore the thing anyway, because the finding arrived in a language none of them was ever taught to read.

It looks like nurses, over and over, holding together with attention and improvisation a system that was designed to work without either.

And, it hides the fact that patients are being harmed in an environment full of harm, because harm is not getting recognized or recorded.

This last one is why I wrote it.

I could have written a shorter and more accusatory book. There is an appetite for books that say the technology is the villain, or that the administrators are, or that the physicians simply need more resilience. Each of those has a large and receptive audience and each of them is wrong in a way that keeps the disease untreated.

The truth is duller and harder. American healthcare has a diagnostic deficiency. It has enormous quantities of information, extraordinary specialists in every domain, and almost no capacity trained to read the whole picture and identify what is wrong. The symptoms have been documented for more than 30 years by people who were predictive and then right. They have never been read together.

Reading symptoms together is a real skill. It is taught in one profession, over a decade, at great expense. And, that profession has been almost entirely absent from the rooms where the decisions get made.

That is the entire book. Everything else is evidence.

I am aware of what I am asking.

I am asking a country to take seriously a role held by people it has never counted, produced at around 50 a year, defined by an institution that admitted a decade ago it did not know what the job was. I am asking boards to fund value that has never been published until now. And, I am asking my own discipline to admit, out loud, that we built a credential before we built the thing it was supposed to certify.

But I have also seen what happens when it works. I have watched an organization go from override rates that were never measured to a governed library, from a documentation burden that was driving people out to notes that get signed before dinner, from data no team trusted to numbers a board could actually use. It is not dramatic while it is happening. It looks like a series of unglamorous meetings, held on schedule, by people with the authority to change something.

Then one day a patient with diabetes has a better visit than they have had in years, and never knows why.

That is the whole of it.

I do not know whether this book will change anything. Many books do not. The evidence in it is real, the arithmetic holds, and neither of those has ever been sufficient on its own, as seventeen chapters just finished demonstrating.

What I know is that the case has now been presented. The history is taken. The examination is documented. The differential is written down and the alternatives are named. The diagnosis is committed. The treatment is specified, priced, and dosed. The prognosis is honest. The referral is written.

The chart is complete.

For those of you who lasted to this point, you have more than enough information to do something about the patient.

THE FIRST STEP IN RECOVERY FROM A DISEASE IS THE RECOGNITION that you have one.

Thomas Powell, MD MS

APPENDIX

APPENDIX A

THE SIX CONDITIONS FOR THE DIAGNOSTICIAN

This is a ten-minute screen. Whether the Diagnostician and informatics is present and functioning at your organization, or only the title.

Chapter 8 established six conditions for Diagnostician and informatics success: education and preparation, funding, positioning, empowering, staffing, employing. This is the instrument for scoring your own organization against them.

When to use it. Do it first today, there is a lot of work to do. Before a search. Before a budget cycle. Before chartering an informatics program. Before believing a vendor. Before a skeptical board or CEO.

How to score. Each condition is met (worth 1 point) or it is not (worth 0). Every condition has a verification test, and the test is what gets scored. The maximum total is six.

Condition 1. Education and Preparation

Has the candidate applying for the CMIO seat been trained in this discipline, and has the executive team been prepared to work with them?

Verification. Name the training: a fellowship, a graduate program, or a structured mentored pathway with defined competencies. Then ask them, without notes, to describe what this role is for.

The correct formal education, while critical, is not necessarily a deal breaker in a field this inconsistently staffed. This book discusses the competencies. They should be aware of them and competent in them. If the first answer is "years of experience with our EHR," and the second is "has a job title," consider marking not met.

Met □ Not met □

Condition 2. Funding

Are the Diagnostician's position and workforce funded as an ongoing strategic commitment, or something less?

Verification. Find the program line in the budget and strategic plan, not the salary line for the Name: the leader, the informatics staff, the governance support, the analytic capacity.

If the program exists only as fractions of other people's time, borrowed from departments reporting elsewhere, not met.

Met □ Not met □

Condition 3. Positioning

Does the seat sit where the decisions are made in full executive standing?

Verification. Trace the reporting line until it reaches a role that can move money. Then check the standing voting list for the meetings where technology, capital, and clinical strategy are decided.

A role reporting into the technology function, or attending meetings by "guest," is only positioned as an adviser. Advisers do not have accountability, and this condition is about accountability.

Met □ Not met □

Condition 4. Empowering

Can this person stop an organization decision?

Verification. Name the last decision they stopped: an alert retired, a build halted, an AI model suspended, a go-live delayed, a vendor commitment refused. Give an example.

Every other condition can be answered aspirationally. This one requires an event that either happened or did not. If the room goes quiet, not met.

Met □ Not met □

Condition 5. Staffing

Does the Diagnostician have a functioning support staff, or only the Name?

Verification. Count the direct reports and the dedicated institutional support to the function. Ask how many of those are borrowed resources from someone else's staff and agenda.

A leader with no staff is a committee invitee with a name.

Met □ Not met □

Condition 6. Employing

Is the Diagnostician fully dedicated to the executive job, or is it a side job within another agenda?

Verification. Read the organizational chart, protected time, and the detailed job description. The signs of a role that is a named placeholder are easy to see.

Chapter 9 argued against the part-time commitment. Keeping your clinical skills current is admirable. A common myth is that working physicians discredit roles not on the "front line." The reality is that a part-time executive signals a part-time effort. There is a lot of work to do.

Met □ Not met □

. . .

TOTAL CONDITIONS MET: _____ / 6

READING THE SCORE

Five or Six. The Name and Form are fully present for Functional success. Your work is maintenance, succession, and the conditions in Chapter 16 that are still arriving. Turn to the Discharge Plan and read "When to Call Back."

Three or Four. The Name exists in some Form, but the Function is constrained by the organization's commitment to the role. Name the missing conditions specifically and fix those, rather than reorganizing anything. One missing condition is usually one conversation with one person.

One or Two. The Name is present and the Form and Function are not. There will not be enough money, time, or resources to make this successful. This is a very common score in American healthcare. Nothing improves until the missing conditions are met, and no additional technology purchase substitutes for meeting them.

Zero. The organization has not even decided to have this Name. That is a legitimate decision that should be made deliberately, in a room, with the financial exposure from Appendix C on the table.

APPENDIX B

THE COMPREHENSIVE INFORMATICS ASSESSMENT

THE FULL EXAMINATION. THIRTY-FIVE ITEMS ACROSS SEVEN DOMAINS.

Relationship to Appendix A. Appendix A asks whether the Diagnostician Function exists. This asks how well it works across the domains it touches. Appendix A takes ten minutes and produces a number. This takes an afternoon and produces a map. Run Appendix A first. If it scores low, the more useful next step is Chapter 16's second door, the mentored pathway and its five expedited interim-coverage arrangements, rather than a deeper assessment.

Purpose. Characterize the diagnostic state of your organization across seven informatics domains, surfacing operative pathology without weeks of audit work.

When to use. Early conversations with executive or board leadership, before beginning a Diagnostician search, before chartering a clinical informatics improvement program, and as a recurring annual self-assessment.

How to score. Circle one number, 1 through 5, for each item. Domain averages suggest the need for intervention. This scorecard is not formally validated. Treat it as a diagnostic conversation tool, not a verdict.

The maturity scale

1 **No capability.** The Function is absent.
2 **Informal or ad hoc.** Informatics activity occurs without structure or accountability.
3 **Partially formalized.** Some structure exists, applied inconsistently.
4 **Formal and consistent.** Chartered, staffed, and applied across the organization.
5 **Mature and continuously improving.** Measured, governed, and refined on a defined cadence.

Domain 1. Cross-Executive Synthesis

Whether the organization has chartered a cross-executive Function with the authority and cadence to drive informatics decisions.

1 Is there a chartered Diagnostician Function with an executive-physician leadership chair?

Score: 1 2 3 4 5

2 Does the executive-physician chair have explicit authority to approve decisions up to a defined dollar or scope threshold without further sign-off?

Score: 1 2 3 4 5

3 Do the chief executive, medical, information, financial, operating, and nursing officers regularly attend as peers?

Score: 1 2 3 4 5

4 Does each meeting produce a decision document?

Score: 1 2 3 4 5

5 Is the cross-disciplinary and cross-executive synthesis Function permanent, with a succession plan?

Score: 1 2 3 4 5

Domain 1 total: _____ / 25

Domain 2. Clinical Informatics Capacity

Whether the organization has staffed clinical informatics at the level required to govern the EHR work, the analytics, and the program's measured financial impact.

1 Is the Diagnostician a full-time C-suite executive?

Score: 1 2 3 4 5

2 Are there compensated physician informaticists and subject matter experts with protected time?

Score: 1 2 3 4 5

3 Is the clinical EHR IT analyst pipeline staffed to the volume of governed change?

Score: 1 2 3 4 5

4 Is the internal pipeline for analysts and informaticist development staffed and funded?

Score: 1 2 3 4 5

5 Is the program's financial impact measured and reported?

Score: 1 2 3 4 5

Domain 2 total: _____ / 25

Domain 3. Decision Support Governance

Whether clinical decision support is governed as a clinical resource, with deactivation authority, override surveillance, consequence ranking, and a defined evidence-review cadence.

1 Is there a chartered clinical decision support committee with deactivation authority and IT analyst prioritization?

Score: 1 2 3 4 5

2 Is the alert override rate measured by alert and in aggregate?

Score: 1 2 3 4 5

3 Has the alert library been ranked by clinical consequence, separating alerts that stand between a physician and a potentially fatal order from those that do not?

Score: 1 2 3 4 5

4 Are alerts delivered as recommendations rather than assessments?

Score: 1 2 3 4 5

5 Are safety events mapped back to decision support coverage on a regular cadence?

Score: 1 2 3 4 5

Domain 3 total: _____ / 25

Domain 4. Artificial Intelligence Oversight

Whether clinical AI model deployment is governed with discontinuation authority, local validation, subgroup monitoring, fabrication auditing, and named clinical ownership.

1 Is there a chartered AI governance body with a physician chair and explicit authority to suspend a deployment?

Score: 1 2 3 4 5

2 Has every deployed AI model been locally validated against your own patient population, with the validation producible on request?

Score: 1 2 3 4 5

3 Is subgroup performance monitored across race, ethnicity, language, age, sex, payer, and geography?

Score: 1 2 3 4 5

4 Are generative and ambient documentation tools audited for fabrication, with the audited rate trended over time?

Score: 1 2 3 4 5

5 Does every deployed AI model have written discontinuation criteria and a named physician who can trigger them?

Score: 1 2 3 4 5

Domain 4 total: _____ / 25

Domain 5. Data Quality and Master Data

Whether the master data layer, the patient index, the terminologies, the lineage, and the dashboard definitions are owned, governed, and surveilled.

1 Is there a named owner for the organization's data quality?

Score: 1 2 3 4 5

2 Are terminology services governed and standardized?

Score: 1 2 3 4 5

3 Is data lineage audited end to end on key organization metrics?

Score: 1 2 3 4 5

4 Are dashboard definitions stable across departments?

Score: 1 2 3 4 5

5 Are your data used to train or tune deployed AI models documented, and are they matched to your own population?

Score: 1 2 3 4 5

Domain 5 total: _____ / 25

Domain 6. Patient Safety and Iatrogenic Surveillance

Whether clinical informatics-impactable safety events are mapped, safeguards are in place, and iatrogenic technology harm (Chapter 6) is tracked as a discrete category.

1 Are reportable safety events mapped to informatics root cause and clinical governance?

Score: 1 2 3 4 5

2 Are patient verification forcing functions in place at order entry and care delivery?

Score: 1 2 3 4 5

3 Is closed-loop result management documented and audited?

Score: 1 2 3 4 5

4 Are decision support, order set, and document template changes governed before deployment?

Score: 1 2 3 4 5

5 Is iatrogenic technology harm (Chapter 6) tracked as a discrete category?

Score: 1 2 3 4 5

Domain 6 total: _____ / 25

Domain 7. Documentation, Workflow, and Burnout

Whether documentation templates and ambient tools are governed, clinician time and usability are measured, and workforce strategy is linked to technology strategy.

1 Are all clinical documentation strategies governed by clinical and informatics leadership?

Score: 1 2 3 4 5

2 Are ambient documentation tools deployed with structured review and signature before filing?

Score: 1 2 3 4 5

3 Is clinician documentation time measured and reported?

Score: 1 2 3 4 5

4 Is clinician usability surveyed and measured annually with a validated instrument?

Score: 1 2 3 4 5

5 Is workforce strategy linked to technology strategy at the executive level?

Score: 1 2 3 4 5

Domain 7 total: _____ / 25

Scoring summary

Add the 7 Domain Totals. Maximum total is 175.

TOTAL: _____ / 175 = _____%

READING THE MAP. DOMAIN AVERAGES, NOT THE TOTAL, ARE WHAT TO act on. A domain averaging below 3 is the operative pathology. 2 or more domains below 3 usually mean an organizational governance problem rather than an isolated domain problem, and the differential should be reworked at the level of Appendix A. The total is useful for one thing only: tracking the same organization against itself, year over year, using the measure.

APPENDIX C

THE BURDEN OF HEALTHCARE'S FINANCIAL DISEASE CALCULATOR

This will help calculate what the absence of a Diagnostician and robust informatics Function is costing your organization, per year, in opportunity. *This tool is physician math extracted from legitimate sources, and argued in Chapter 8.*

Chapter 8 assembled the national financial exposure and converted it to a per-physician figure. This is the worksheet applying to your own organization.

One thing before any figures. A financial exposure is the size of a category where loss occurs. It is not a claim that the category is fully recoverable, and nothing produced by this worksheet should be presented as recoverable savings. It is recoverable opportunity. That distinction is the difference between a defensible business case and a losing argument.

Every national figure below is taken at the low end of its published range. Pricing failure and fraud are excluded.

Step one. Your estimated financial exposure

Your physician count. Physicians employed and contracted. Multiply by the per-physician financial exposure established in Chapter 8.

_______ physicians
× $840,304
= $______ annual financial exposure

What is inside that multiplication. Listed for the reader who wants to know. Do not add these to your own figures. The per-physician number already contains them.

- Paperwork that wastes time: $265.6 billion. Prior authorization, denied claims, quality reporting.
- The half-day that doesn't count: $180 billion. Physician time spent on the EHR.
- Harm, and what follows it: $112 billion. Preventable harm, liability, documentation failure.
- Care we did not need to give: $103 billion. Repeated tests, low-value treatment.
- Technology we pay for and do not use: some percentage of healthcare's $69 billion annual technology spend. Unused licenses, redundant applications.
- Money we earned and did not collect: $35 billion. Documentation integrity, charge capture.
- Losing the people: $30 billion. Turnover, agency premium, burnout.
- Other measured financial exposures: $72.9 billion. Downtime, avoidable days, regulatory exposure.
- Total: $867.5 billion

Step two. Your estimated program cost

What a constituted Diagnostician and informatics Function costs at your scale. Chapter 8 priced a 1,000-physician program at roughly $2.84 million a year ($2,840/physician), with 9 of 11 of that staff already employed somewhere under other leaders, leaving roughly $1,184 per physician.

Your Total Program Cost: $2,840/physician ×
# of physicians employed______
= $_______ **Total Program Cost**

Your New Money Required: $1,184/physician ×
# of physicians employed______

= $_______ **New money required**

Step three. Your breakeven

New money divided by financial exposure.

$_____ New Money
÷
$_____ Financial Exposure
× 100
=
______ % per $100 break-even

Chapter 8's model produces 0.14%, which is $0.14 (new money) of every $100. This is the number representing opportunity, not the size of the return.

Step four. The three tests

Run these against your own numbers before presenting any of it. They are built from Chapter 8's arithmetic.

Test one. Breakeven. Computed above. Lead with it.

Test two. Robustness. Delete your largest category and recompute. Then delete every line of shaky estimates. This model does not need every line for the argument and it was already at the low end.

Survives deletion of the largest category?

Yes / No

Survives deletion of all estimated lines?

Yes / No

Test three. Margin multiplier. Take your operating margin and divide 1 by it. That is what a dollar of avoided cost contributes relative to a dollar of new revenue.

1 ÷ ______ your margin as a decimal
= ______

At a 3% margin, the multiplier is 33. Use your own margin, not a national one. State the claim precisely: that at your margin, $1 not spent contributes what every $33 earned contributes.

Step five. What is curable and what is only treatable

Roughly a quarter of the available return comes from categories that can be cured (stopped) rather than only treated (shrunk).

Curable. Denials on claims that were pre-approved. Quality measures computable from discrete fields. Duplicate orders inside your own system. Unused licenses.

Treatable only. Everything else. Harm can be reduced substantially and cannot be eliminated. Documentation burden shrinks and does not disappear.

Some diseases you can only help. Some you can cure.

Step six. The timeline

This is the largest variable, based upon local environment. Full organization maturity runs about a decade. Kaushal's institution did not turn a net profit until year six. Many will show substantial impact long before. An organization evaluating this on a short-term horizon will worry that it doesn't pay for itself yet. That worry is premature.

APPENDIX D

THE AI MODEL INVENTORY

One form per deployed AI model. If you cannot fill it in, you have your finding.

Purpose. Chapter 13 established that essentially no American health system can produce local validation, subgroup performance, and discontinuation criteria for the AI models already running in its building. This is the form.

When to use. For every artificial intelligence or predictive model currently in production, before approving any new one, and as the first artifact of any governance body chartered under Chapter 13.

How to use it. One form per model. A form that cannot be completed is not an administrative gap. It is the finding, and it should be reported to the governance body as such.

AI model identification

Name and version _________
Vendor ___________

Date of go-live __________
Last reviewed __________
Clinical owner _________
If this line is blank, stop here. A tool not defended by name has no clinical sponsor.

Indication

What clinical or operational problem is this AI model for?
What is it explicitly not for?
Who acts on its output, and what are they expected to do?

What the authorization certifies

Regulatory pathway, if any: ___________

Note for context: approximately 96.5% of authorized artificial intelligence devices reached market through the 510(k) pathway, which asks for equivalence to a predicate rather than a trial. Fewer than 1 in 5 referenced prospective validation data (Singh et al., 2025; Loganathan et al., 2026). Clearance is not evidence of performance in your population.

Vendor-reported performance

Validated against our own patients?
Yes / No
Date, sample, and method: ___________
Measured performance (sensitivity): _______
Difference from sensitivity ___________%

If measured performance falls materially below what was represented, that finding goes to the vendor in writing. Wong's paper on a widely deployed sepsis model exists because one academic center did this once (Wong et al., 2021).

Subgroup performance

Examined by: Race ☐ Ethnicity ☐ Language ☐ Age ☐ Sex ☐
Payer ☐ Geography ☐
Findings ________________
Date of most recent review ____
If no box is checked, note that Obermeyer's algorithm passed every test that was run (Obermeyer et al., 2019).

Monitoring

Metrics tracked __________
Review cadence __________
Reviewed by ____________
Drift detection method ______
For generative or ambient tools, audited fabrication rate
_______%
Trend: Falling / Flat / Rising / Not measured

Discontinuation criteria

Written in advance, this AI model will be suspended if:
1
2
3
Who holds the authority to suspend it?
____

A model that cannot be stopped by a named person is not governed. It is installed.

Governance log

Approved by ___________

Date __________

Next scheduled review ______

Prior suspensions or modifications: ___________

Reading the completed inventory

Lay the forms side by side and count.

How many have a named clinical owner? Anything without one is unsponsored.

How many have local validation? Anything without it was just installed rather than validated.

How many have subgroup performance? This count will be the lowest, and it has the clearest published harm behind it.

How many have discontinuation criteria and a named person who can trigger them? Chapter 12's test: a program that only adds, accumulates.

AND, HOW MANY ANSWERS, OR EVEN ENTIRE FORMS, ARE INCOMPLETE? That will be a telling number.

GLOSSARY OF ACRONYMS

All acronyms are spelled out at their first use in the text; this glossary is the comprehensive reference for any term encountered without a nearby expansion. My coined terms and the framework they belong to are defined in "A Note on Method and Structure."

- AAMC. Association of American Medical Colleges.
- ABPM. American Board of Preventive Medicine.
- ACDIS. Association of Clinical Documentation Integrity Specialists.
- ACGME. Accreditation Council for Graduate Medical Education. Accredits residency and fellowship training programs in the United States.
- ADE. Adverse drug event.
- AI. Artificial intelligence.
- AMA. American Medical Association.
- AMIA. American Medical Informatics Association.
- ASTP. Assistant Secretary for Technology Policy. The federal office that absorbed ONC's functions.
- BID. Bis in die. Twice daily (medication-dosing notation).
- CDI. Clinical documentation improvement (also called clinical documentation integrity). The discipline of ensuring clinical records are complete, accurate, and reflect the patient's true clinical picture.
- CDS. Clinical decision support. Computerized tools that assist clinicians in making evidence-based decisions at the point of care. Also known as computerized clinical decision support.
- CEO. Chief Executive Officer.
- CFO. Chief Financial Officer.
- CHRO. Chief Human Resources Officer.
- CIO. Chief Information Officer.
- CKD. Chronic kidney disease.
- CMIO. Chief Medical Information Officer. The executive-physician of this book; the role uniquely qualified to lead the diagnostic and treatment process of the healthcare organization.
- CMO. Chief Medical Officer.
- CMS. Centers for Medicare & Medicaid Services. The federal agency administering Medicare and Medicaid.

- **CNO**. Chief Nursing Officer.
- **COO**. Chief Operating Officer.
- **CPOE**. Computerized provider order entry. Electronic ordering of medications, tests, and procedures.
- **CPT**. Current Procedural Terminology. AMA-maintained coding system for medical procedures.
- **CT**. Computed tomography.
- **DICOM**. Digital Imaging and Communications in Medicine. Standard for medical imaging data and transfer.
- **DMAIC**. Define, Measure, Analyze, Improve, Control. Six Sigma improvement framework.
- **DRG**. Diagnosis-related group. Patient classification system used for inpatient reimbursement.
- **ECRI**. ECRI Institute. Independent nonprofit safety-research organization.
- **ED**. Emergency department.
- **EHR**. Electronic health record.
- **EKG**. Electrocardiogram.
- **EVS**. Environmental services. Hospital cleaning and sanitation operations.
- **FDA**. U.S. Food and Drug Administration.
- **FHIR**. Fast Healthcare Interoperability Resources. Modern healthcare data-exchange standard. See HL7 FHIR.
- **GDP**. Gross domestic product.
- **GLP**. Glucagon-like peptide. As in GLP-1 receptor agonists used in diabetes and obesity.
- **HHS**. U.S. Department of Health and Human Services.
- **HIMSS**. Healthcare Information and Management Systems Society.
- **HIPAA**. Health Insurance Portability and Accountability Act of 1996.
- **HITECH**. Health Information Technology for Economic and Clinical Health Act of 2009.
- **HL7**. Health Level Seven. Standards-development organization for healthcare data.
- **HL7 FHIR**. Health Level Seven Fast Healthcare Interoperability Resources.
- **HRSA**. Health Resources and Services Administration.
- **ICD**. International Classification of Diseases. WHO-maintained clinical coding system.
- **ICU**. Intensive care unit.

- **IOM**. Institute of Medicine. Now the National Academy of Medicine (NAM).
- **I-PASS**. Illness severity, Patient summary, Action list, Situational awareness, Synthesis. Structured handoff system.
- **IT**. Information Technology.
- **KFF**. Formerly the Kaiser Family Foundation. Health policy research organization.
- **KLAS**. KLAS Research. Independent firm that surveys healthcare technology performance and user satisfaction.
- **LOINC**. Logical Observation Identifiers Names and Codes. Standardized terminology for laboratory and clinical observations.
- **LOS**. Length of stay (hospital).
- **MACRA**. Medicare Access and CHIP Reauthorization Act of 2015.
- **MD**. Doctor of Medicine.
- **MIPS**. Merit-based Incentive Payment System.
- **MRI**. Magnetic resonance imaging.
- **MS**. Master of Science.
- **NASEM**. National Academies of Sciences, Engineering, and Medicine.
- **NLM**. National Library of Medicine.
- **NLP**. Natural language processing.
- **NSI**. NSI Nursing Solutions, Inc. Source of national nurse-staffing turnover data.
- **ONC**. Office of the National Coordinator for Health Information Technology.
- **PACS**. Picture Archiving and Communication System. Storage and retrieval system for medical images.
- **PCS**. Procedure Coding System (used in ICD-10-PCS).
- **PDSA**. Plan, Do, Study, Act. Quality-improvement cycle.
- **RAND**. RAND Corporation. Nonprofit research organization.
- **RN**. Registered Nurse.
- **ROI**. Return on investment.
- **RxNORM**. a standardized nomenclature for clinical drugs, maintained by the U.S. National Library of Medicine.
- **SAFER**. Safety Assurance Factors for EHR Resilience. ONC self-assessment guides for EHR safety.
- **SMART**. Substitutable Medical Applications and Reusable Technologies. App platform built on FHIR.
- **SNOMED**. Systematized Nomenclature of Medicine, Clinical Terms. Comprehensive clinical terminology.

- **TEFCA.** Trusted Exchange Framework and Common Agreement. National framework for health information exchange.
- **TJC.** The Joint Commission.
- **UMLS.** Unified Medical Language System. NLM-maintained set of files and software linking health and biomedical vocabularies.
- **VA.** U.S. Department of Veterans Affairs.
- **VBP.** Value-based purchasing.
- **WHO.** World Health Organization.

BIBLIOGRAPHY

Cited works are listed alphabetically by first author. Where a landmark paper has been confirmed or extended by more recent work, both are listed and both are cited in the text. Citations in the text use author-date format.

AAMC (Association of American Medical Colleges). The Complexities of Physician Supply and Demand: Projections from 2021 to 2036. Washington, DC: AAMC; 2024.

ABPM (American Board of Preventive Medicine). Clinical Informatics certification requirements. The practice pathway closed following the 2025 examination cycle, with qualifying practice experience counted through June 30, 2025. Accessed 2026.

ACDIS (Association of Clinical Documentation Integrity Specialists). 2024 CDI Industry Survey. East Greenwich, RI: ACDIS; 2024.

ACGME (Accreditation Council for Graduate Medical Education). Data Resource Book, Academic Year 2024-2025. Chicago, IL: ACGME; 2025.

Adelman JS, Applebaum JR, Schechter CB, et al. Effect of restriction of the number of concurrently open records in an electronic health record on wrong-patient order errors: a randomized clinical trial. *JAMA*. 2019;321(18):1780-1787.

Adelman JS, Kalkut GE, Schechter CB, et al. Understanding and preventing wrong-patient electronic orders: a randomized controlled trial. J Am Med Inform Assoc. 2013;20(2):305-310.

Adler-Milstein J, Daniel G, Grossmann C, et al. Return on Information: A Standard Model for Assessing Institutional Return on Electronic Health Records. NAM Perspectives. Washington, DC: National Academy of Medicine; 2014.

Adler-Milstein J, Holmgren AJ, Kralovec P, et al. EHR adoption in US hospitals: the emergence of a digital "advanced use" divide. J Am Med Inform Assoc. 2017;24(6):1142-1148.

AHA (American Hospital Association). 2025. The Cost of Caring Report. Chicago: AHA; 2026.

AHA (American Hospital Association). The Current State of Hospital Finances: Fall 2022 Update. Chicago: AHA; September 2022.

AHA (American Hospital Association). Regulatory Overload: Assessing the Regulatory Burden on Health Systems, Hospitals and Post-acute Care Providers. Prepared by Manatt Health. Washington, DC: American Hospital Association; 2017.

AMA (American Medical Association). 2025 AMA Prior Authorization Physician Survey. Chicago, IL: American Medical Association; 2025.

AMA (American Medical Association). Physician Burnout, Stress, and Turnover Report. Chicago, IL: AMA; 2024.

Appari A, Johnson ME, Anthony DL. Meaningful use of EHR systems and process quality of care: evidence from a panel data analysis of U.S. acute-care hospitals. Health Serv Res. 2013;48(2 Pt 1):354-375.

Asgari E, Montaña-Brown N, Dubois M, et al. A framework to assess clinical safety and hallucination rates of LLMs for medical text summarisation. npj Digit Med. 2025;8:274.

Ash JS, Berg M, Coiera E. Some unintended consequences of IT in health care: the nature of patient care information system-related errors. J Am Med Inform Assoc. 2004;11(2):104-112.

Asplin BR, Magid DJ, Rhodes KV, et al. A conceptual model of ED crowding. Ann Emerg Med. 2003;42(2):173-180.

Balas EA, Boren SA. Managing clinical knowledge for health care improvement. Yearb Med Inform. 2000;9(1):65-70.

Bates DW, Leape LL, Cullen DJ, et al. Effect of computerized physician order entry and a team intervention on prevention of serious medication errors. JAMA. 1998;280(15):1311-1316.

Beauvais B, Kruse CS, Fulton L, et al. Association of electronic health record vendors with hospital financial and quality performance: retrospective data analysis. J Med Internet Res. 2021;23(4):e23961.

Bevan G, Hood C. Have targets improved performance in the English NHS? BMJ. 2006;332(7538):419-422.

Bloom N, Han R, Liang J. Hybrid working from home improves retention without damaging performance. Nature. 2024;630(8018):920-925.

Blumenthal D. Launching HITECH. N Engl J Med. 2010;362(5):382-385.

Bright TJ, Wong A, Dhurjati R, et al. Effect of clinical decision-support systems: a systematic review. Ann Intern Med. 2012;157(1):29-43.

Brookings Institution, Hutchins Center on Fiscal and Monetary Policy. The Hutchins Center Explains: Prescription Drug Spending. Washington, DC: Brookings Institution; 2022.

Buckhead FMV. CMIO compensation in nonprofit hospitals: A longitudinal study. BFMV Clinical Leadership Valuation Series. 2024; Accessed August 25, 2026. buckheadfmv.com.

Budzyń K, Romańczyk M, Kitala D, et al. Endoscopist deskilling risk after exposure to artificial intelligence in colonoscopy: a multicentre, observational study. The Lancet Gastroenterology & Hepatology. 2025;10(10):896-903.

Buntin MB, Jain SH, Blumenthal D. Health Information Technology: Laying the Infrastructure for National Health Reform. Health Affairs. 2010;29(6):1214-1219.

Burwell SM. Setting value-based payment goals - HHS efforts to improve U.S. health care. N Engl J Med. 2015;372(10):897-899.

Campbell DT. Assessing the impact of planned social change. Eval Program Plann. 1979;2(1):67-90.

Candello. For the Record: The Effect of Documentation on Defensibility and Patient Safety. Boston, MA: The Risk Management Foundation of the Harvard Medical Institutions; 2024.

Casalino LP, Gans D, Weber R, et al. US physician practices spend more than $15.4 billion annually to report quality measures. Health Aff (Millwood). 2016;35(3):401-406.

Centers for Medicare & Medicaid Services (CMS). Promoting Interoperability Programs and Hospital VBP. Baltimore, MD: CMS; 2024.

Char DS, Abramoff MD, Feudtner C. Identifying ethical considerations for machine learning healthcare applications. Am J Bioeth. 2020;20(11):7-17.

Char DS, Shah NH, Magnus D. Implementing machine learning in health care - addressing ethical challenges. N Engl J Med. 2018;378(11):981-983.

Cho I, Lee Y, Lee J-H, Bates DW. Wide variation and patterns of physicians' responses to drug-drug interaction alerts. Int J Qual Health Care. 2019;31(2):89-95.

Classen DC, Holmgren AJ, Co Z, Newmark LP, Seger D, Danforth M, Bates DW. National trends in the safety performance of electronic health record systems from 2009 to 2018. JAMA Netw Open. 2020;3(5):e205547.

CMS (Centers for Medicare & Medicaid Services). Medicare Fee-for-Service Recovery Audit Program. Baltimore, MD: CMS; 2025a.

CMS (Centers for Medicare & Medicaid Services). National Health Expenditure Data, 2024. CMS Office of the Actuary; 2025b.

CMS (Centers for Medicare & Medicaid Services). National Health Expenditures 2024 Highlights. Baltimore, MD: CMS; 2025c.

Coiera E, Chan A, Brooke-Cowden K, et al. Clinical and economic impact of digital dashboards on hospital inpatient care: a systematic review. JAMIA Open. 2025;8(4):ooaf078.

Co Z, Holmgren AJ, Classen DC, et al. The tradeoffs between safety and alert fatigue: data from a national evaluation of hospital medication-related clinical decision support. J Am Med Inform Assoc. 2020;27(8):1252-1258.

Congressional Budget Office. Evidence on the costs and benefits of health information technology. Congress of the United States; 2008.

Consolidated Appropriations Act, 2026. Extension of Medicare telehealth flexibilities through December 31, 2027.

Croskerry P. Achieving quality in clinical decision making: cognitive strategies and detection of bias. Acad Emerg Med. 2002;9(11):1184-1204.

Croskerry P. The importance of cognitive errors in diagnosis and strategies to minimize them. Acad Med. 2003;78(8):775-780.

Croskerry P. A universal model of diagnostic reasoning. Acad Med. 2009;84(8):1022-1028.

Cutler DM. Reducing Administrative Costs in U.S. Health Care. Washington, DC: The Hamilton Project, Brookings Institution; March 2020.

Damschroder LJ, Aron DC, Keith RE, et al. Fostering implementation of health services research findings into practice: a consolidated framework for advancing implementation science. Implement Sci. 2009;4:50.

Damschroder LJ, Reardon CM, Widerquist MAO, Lowery JC. The updated Consolidated Framework for Implementation Research based on user feedback. Implement Sci. 2022;17(1):75.

Daneshjou R, Vodrahalli K, Novoa RA, et al. Disparities in dermatology AI performance on a diverse, curated clinical image set. Sci Adv. 2022;8(31):eabq6147.

Definitive Healthcare. Average Hospital IT Expenses and Budget Benchmarks. Framingham, MA: Definitive Healthcare; 2024.

Detmer DE, Munger BS, Lehmann CU. Clinical informatics board specialty certification: a status report on the field. J Am Med Inform Assoc. 2024;31(2):447-453.

Detmer DE, Shortliffe EH. Clinical informatics: prospects for a new medical subspecialty. JAMA. 2014;311(20):2067-2068.

Dougherty M, Seabold S, White SE. CDI programs and the impact on case mix index. J AHIMA. 2013;84(7):60-64.

Eberly LA, Kallan MJ, Julien HM, et al. Patient characteristics associated with telemedicine access for primary and specialty ambulatory care during the COVID-19 pandemic. JAMA Netw Open. 2020;3(12):e2031640.

ECRI Institute. PSO Deep Dive: Patient Identification. Plymouth Meeting, PA: ECRI Institute; 2016.

ECRI Institute. Top 10 Health Technology Hazards for 2025. Plymouth Meeting, PA: ECRI Institute; 2024.

El-Kareh R, Hasan O, Schiff GD. Use of health IT to reduce diagnostic errors. BMJ Qual Saf. 2013;22 Suppl 2:ii40-ii51.

Elstein AS, Shulman LS, Sprafka SA. Medical Problem Solving: An Analysis of Clinical Reasoning. Cambridge, MA: Harvard University Press; 1978.

FDA (Food and Drug Administration). 2026. Artificial Intelligence-Enabled Medical Devices. Silver Spring, MD: FDA; data as of December 2025. Accessed March 2026.

Forrester Research. US Tech Forecast for Healthcare Providers, 2026. Cambridge, MA: Forrester Research; 2026.

Francis R. Report of the Mid Staffordshire NHS Foundation Trust Public Inquiry. London: The Stationery Office; 2013.

Goddard K, Roudsari A, Wyatt JC. Automation bias: a systematic review of frequency, effect mediators, and mitigators. J Am Med Inform Assoc. 2012;19(1):121-127.

Goddard K, Roudsari A, Wyatt JC. Automation bias: empirical results assessing influencing factors. Int J Med Inform. 2014;83(5):368-375.

Gottlieb JD, Polyakova M, Rinz K, et al. Who Values Human Capitalists' Human Capital? The Earnings and Labor Supply of U.S. Physicians. NBER Working Paper No. 31469. Cambridge, MA: National Bureau of Economic Research; July 2023.

Graber ML, Kissam S, Payne VL, et al. Cognitive interventions to reduce diagnostic error: a narrative review. BMJ Qual Saf. 2012;21(7):535-557.

Greenhalgh T, Wherton J, Papoutsi C, et al. Beyond adoption: a new framework for theorizing and evaluating nonadoption, abandonment, and challenges to the scale-up, spread, and sustainability of health and care technologies. J Med Internet Res. 2017;19(11):e367.

Guidehouse. Annual Supply Chain Savings Opportunity Reaches $25.7 Billion for U.S. Hospitals, Guidehouse Analysis Finds. Washington, DC: Guidehouse; 2019.

Gunja MZ, Gumas ED, Williams RD. High U.S. Health Care Spending: Where Is It All Going? New York, NY: The Commonwealth Fund; October 2023.

Hamidi MS, Bohman B, Sandborg C, et al. Estimating institutional physician turnover attributable to self-reported burnout and associated financial burden: a case study. BMC Health Serv Res. 2018;18(1):851.
Hammond WE. The making and adoption of health data standards. Health Affairs. 2005;24(5):1205-1213.
Hammond WE, Bent B, West VL. Goodbye electronic health record? Stud Health Technol Inform. 2022;298:107-111.
Han S, Shanafelt TD, Sinsky CA, et al. Estimating the attributable cost of physician burnout in the United States. Ann Intern Med. 2019;170(11):784-790.
HHS (U.S. Department of Health and Human Services). HHS expands secure access to health records through the TEFCA network, announces milestone of one billion health records exchanged. Press release; 2026.
Health Information Technology for Economic and Clinical Health Act. Enacted as part of the American Recovery and Reinvestment Act of 2009.
HRSA (Health Resources and Services Administration). State of the U.S. Health Care Workforce 2024. Rockville, MD: U.S. Department of Health and Human Services. 2024.
Himmelstein DU, Campbell T, Woolhandler S. Health care administrative costs in the United States and Canada, 2017. Ann Intern Med. 2020;172(2):134-142. doi:10.7326/M19-2818.
HIMSS (Healthcare Information and Management Systems Society). 2019 HIMSS U.S. Leadership & Workforce Survey Report. Chicago, IL: HIMSS; 2019.
HIPAA. Health Insurance Portability and Accountability Act of 1996. Pub. L. No. 104-191, § 1173(b), 110 Stat. 1936. Enacted August 21, 1996.
Hodgson T, Coiera E. Risks and benefits of speech recognition for clinical documentation: a systematic review. J Am Med Inform Assoc. 2016;23(e1):e169-e179.
Holmgren AJ, Apathy NC. Trends in US hospital electronic health record vendor market concentration, 2012-2021. J Gen Intern Med. 2023;38(7):1765-1767.
Holmgren AJ, Apathy NC, Kanter GP. Electronic health record market consolidation and implications for cybersecurity. Health Aff Sch. 2025;3(8):qxaf164.
Holmgren AJ, Fenton CL, Thombley R, et al. Ambient artificial intelligence scribes and physician financial productivity. JAMA Netw Open. 2026;9(1):e2553233.
Holmgren AJ, Kuznetsova M, Classen D, Bates DW. Assessing hospital electronic health record vendor performance across publicly reported quality measures. J Am Med Inform Assoc. 2021;28(10):2101-2107.
IOM (Institute of Medicine). Best Care at Lower Cost: The Path to Continuously Learning Health Care in America. Smith M, Saunders R, Stuckhardt L, McGinnis JM, eds. Washington, DC: National Academies Press; 2012.
IOM (Institute of Medicine), Committee on Improving the Patient Record. The Computer-Based Patient Record: An Essential Technology for Health Care. Dick RS, Steen EB, eds. Washington, DC: National Academy Press; 1991.
IOM (Institute of Medicine). Crossing the Quality Chasm: A New Health System for the 21st Century. Washington, DC: National Academies Press; 2001.
James JT. A new, evidence-based estimate of patient harms associated with hospital

care. J Patient Saf. 2013;9(3):122-128.

Kahn MG, Callahan TJ, Barnard J, Bauck AE, Brown J, et al. A harmonized data quality assessment terminology and framework for the secondary use of electronic health record data. EGEMS (Wash DC). 2016;4(1):1244.

Kahneman D. Thinking, Fast and Slow. New York, NY: Farrar, Straus and Giroux; 2011.

Kaiser Family Foundation (KFF). Hospital Spending Accounted for 40% of the Growth in National Health Spending Between 2022 and 2024. San Francisco, CA: KFF; 2025.

Kalvelage C, Rademacher S, Dohmen S, et al. Decision-making authority during tele-ICU care reduces mortality and length of stay: a systematic review and meta-analysis. Critical Care Medicine. 2021;49(7):1169-1181.

Kane EM, Scheulen JJ, Püttgen A, Efron DT. Impact of a centralized health system operations center on emergency department boarding and inpatient throughput. Journal of Healthcare Management. 2019;64(5):321-333.

Kannry J, Sengstack P, Thyvalikakath TP, et al. The Chief Clinical Informatics Officer (CCIO): AMIA Task Force Report on CCIO Knowledge, Education, and Skillset Requirements. Appl Clin Inform. 2016;7(1):143-176.

Kanter GP, Rekowski JR, Kannarkat JT. Lessons from the Change Healthcare ransomware attack. JAMA Health Forum. 2024;5(9):e242764.

Kaufman Hall. National Hospital Flash Report: December 2025 Metrics. Chicago, IL: Kaufman Hall; 2026.

Kaushal R, Jha AK, Franz C, et al. Return on investment for a computerized physician order entry system. J Am Med Inform Assoc. 2006;13(3):261-266.

Kawamoto K, Houlihan CA, Balas EA, Lobach DF. Improving clinical practice using CDS systems: a systematic review of trials to identify features critical to success. BMJ. 2005;330(7494):765.

Kazley AS, Simpson AN, Simpson KN, Teufel R. Association of electronic health records with cost savings in a national sample. The American Journal of Managed Care. 2014;20(6):e183-e190.

Keehan SP, Fiore JA, Poisal JA, et al. National Health Expenditure Projections, 2024-33: Despite Insurance Coverage Declines, Health to Grow as Share of GDP. Health Aff (Millwood). 2025;44(7):927-937.

Kellermann AL, Jones SS. What it will take to achieve the as-yet-unfulfilled promises of health IT. Health Aff (Millwood). 2013;32(1):63-68.

Kesson EM, Allardice GM, George WD, et al. Effects of multidisciplinary team working on breast cancer survival: retrospective, comparative, interventional cohort study of 13,722 women. BMJ. 2012;344:e2718.

Kilbridge PM, Welebob EM, Classen DC. Development of the Leapfrog methodology for evaluating hospital implemented inpatient computerized physician order entry systems. Qual Saf Health Care. 2006;15(2):81-84.

Kim E, Van Cain M, Hron J. Survey of clinical informatics fellows graduating 2016-2024: experiences before and during fellowship. J Am Med Inform Assoc. 2023;30(10):1608-1613.

KLAS Research. Arch Collaborative: Reducing Nursing Documentation Burden 2025. Orem, UT: KLAS Research; 2025.

KLAS Research; 2026a. Clinician EHR Experience 2026. Arch Report. KLAS Research; Accessed July 28, 2026.

KLAS Research; 2026b. US Acute Care EHR Market Share 2026. Orem, UT: KLAS Research; 2026.

Kohn LT, Corrigan JM, Donaldson MS, eds. To Err Is Human: Building a Safer Health System. Washington, DC: National Academies Press; 1999.

Koppel R, Metlay JP, Cohen A, et al. Role of computerized physician order entry systems in facilitating medication errors. JAMA. 2005;293(10):1197-1203.

Koppel R, Wetterneck T, Telles JL, Karsh BT. Workarounds to barcode medication administration systems: their occurrences, causes, and threats to patient safety. J Am Med Inform Assoc. 2008;15(4):408-423.

Kreimeyer K, Foster M, Pandey A, et al. Natural language processing systems for capturing and standardizing unstructured clinical information: a systematic review. J Biomed Inform. 2017;73:14-29.

Kripalani S, LeFevre F, Phillips CO, et al. Deficits in communication and information transfer between hospital-based and primary care physicians: implications for patient safety and continuity of care. JAMA. 2007;297(8):831-841.

Kwan JL, Lo L, Ferguson J, et al. Computerised CDS systems and absolute improvements in care: meta-analysis of controlled clinical trials. BMJ. 2020;370:m3216.

Lammers EJ, Adler-Milstein J, Kocher KE. Does health information exchange reduce redundant imaging? Evidence from emergency departments. Med Care. 2014;52(3):227-234.

The Leapfrog Group. CPOE Evaluation Tool. Washington, DC: The Leapfrog Group; 2025.

Ledley RS, Lusted LB. Reasoning foundations of medical diagnosis: symbolic logic, probability, and value theory aid our understanding of how physicians reason. Science. 1959;130(3366):9-21.

Lee B, Kim K, Choi JY, et al. Efficacy of the multidisciplinary tumor board conference in gynecologic oncology: a prospective study. Medicine (Baltimore). 2017;96(48):e8089.

Lehmann CU, Gundlapalli AV, Williamson JJ, et al. Five years of clinical informatics board certification for physicians in the United States of America. Yearb Med Inform. 2018a;27(1):237-242.

Lehmann CU, Longhurst CA, Hersh W, et al. Clinical informatics fellowship programs: in search of a viable financial model. Appl Clin Inform. 2018b;9(2):265-267.

Levick DL, Stern G, Meyerhoefer CD, et al. Reducing unnecessary testing in a CPOE system through implementation of a targeted CDS intervention. BMC Medical Informatics and Decision Making. 2013;13:43.

Lewis AE, Weiskopf N, Abrams ZB, et al. Electronic health record data quality assessment and tools: a systematic review. J Am Med Inform Assoc. 2023;30(10):1730-1740.

Li KY, Ngai KM, Genes N. Differences in antibiotic prescribing rates for telemedicine encounters for acute respiratory infections. Journal of Telemedicine and Telecare. 2024;30(3):525-530.

Lindberg DAB, Humphreys BL, McCray AT. The Unified Medical Language System. Methods of Information in Medicine. 1993;32(4):281-291.

Loganathan A, Friedman M, Waseem T, et al. Current state of Food and Drug Administration-approved artificial intelligence/machine learning medical devices: pathways, transparency, and evidence gaps. Journal of Medical Artificial Intelligence. 2026;9:38.

Low S, Belcher E, Jones S. Matching identifiers in electronic health records: Implications for duplicate records and patient safety. Journal of Innovation in Health Informatics. 2013;20(4):219-225.

Lyell D, Coiera E. Automation bias and verification complexity: a systematic review. J Am Med Inform Assoc. 2017;24(2):423-431.

Makary MA, Daniel M. Medical error - the third leading cause of death in the US. BMJ. 2016;353:i2139.

Mandel JC, Kreda DA, Mandl KD, et al. SMART on FHIR: a standards-based, interoperable apps platform for EHR. J Am Med Inform Assoc. 2016;23(5):899-908.

Mandl KD, Kohane IS. Escaping the EHR trap-the future of health IT. N Engl J Med. 2012;366(24):2240-2242.

McDermott KW, Jiang HJ. Characteristics and Costs of Potentially Preventable Inpatient Stays, 2017. HCUP Statistical Brief #259. Rockville, MD: Agency for Healthcare Research and Quality; 2020.

McGlynn EA, Asch SM, Adams J, et al. The quality of health care delivered to adults in the United States. N Engl J Med. 2003;348(26):2635-2645.

Mebrahtu TF, Lawrence S, Skyrme S. Do operational and clinical benefits automatically follow electronic health record adoption? A systematic overview of implementation hurdles. International Journal of Medical Informatics. 2023;170:104952.

MedPAC, Payment Basics: Hospital Acute Inpatient Services Payment System, Washington, DC. 2021.

Mello MM, Chandra A, Gawande AA, Studdert DM. National costs of the medical liability system. Health Aff (Millwood). 2010;29(9):1569-1577.

Melnick ER, Dyrbye LN, Sinsky CA, et al. The association between perceived EHR usability and professional burnout among US physicians. Mayo Clin Proc. 2020;95(3):476-487.

Merritt Hawkins. 2019 Physician Inpatient/Outpatient Revenue Survey. Irving, TX: Merritt Hawkins; 2019.

Metzger J, Welebob E, Bates DW, et al. Mixed results in the safety performance of computerized physician order entry. Health Aff (Millwood). 2010;29(4):655-663.

Miake-Lye IM, Cogan AM, Mak S, et al. Transitioning from one electronic health record to another: a systematic review. J Gen Intern Med. 2023;38(13):2940-2949.

Morley C, Unwin M, Peterson GM, et al. Emergency department crowding: a systematic review of causes, consequences and solutions. PLoS One. 2018;13(8):e0203316.

Morris ZS, Wooding S, Grant J. The answer is 17 years, what is the question: understanding time lags in translational research. J R Soc Med. 2011;104(12):510-520.

Muehlematter UJ, Bluethgen C, Vokinger KN. FDA-cleared artificial intelligence and machine learning-based medical devices and their 510(k) predicate networks. The Lancet Digital Health. 2023;5(9):e618-e626.

Muller JZ. The Tyranny of Metrics. Princeton, NJ: Princeton University Press; 2018.

Murphy DR, Meyer AND, Sittig DF, et al. Application of electronic algorithms to improve diagnostic evaluation for bladder cancer in primary care. Am J Med. 2017;130(8):906-913.

National Academies of Sciences, Engineering, and Medicine (NASEM). Improving Diagnosis in Health Care. Washington, DC: National Academies Press; 2015.

National Academies of Sciences, Engineering, and Medicine (NASEM). Taking Action Against Clinician Burnout: A Systems Approach to Professional Well-Being. Washington, DC: National Academies Press; 2019.

National Library of Medicine. Minutes of the Board of Regents; 2018, listing the history and sunset of the F37 and F38 individual informatics fellowship programs.

Newman-Toker DE, Nassery N, Schaffer AC, et al. Burden of serious harms from diagnostic error in the USA. BMJ Qual Saf. 2024;33(2):109-120.

NSI Nursing Solutions. 2026 NSI National Health Care Retention & RN Staffing Report. East Petersburg, PA: NSI Nursing Solutions; 2026.

Obermeyer Z, Powers B, Vogeli C, Mullainathan S. Dissecting racial bias in an algorithm used to manage the health of populations. Science. 2019;366(6464):447-453.

Olsen L, Aisner D, McGinnis JM, eds. The Learning Healthcare System: Workshop Summary. Institute of Medicine Roundtable on Evidence-Based Medicine. Washington, DC: National Academies Press; 2007.

Olson KD, Meeker D, Troup E, et al. Use of ambient AI scribes to reduce administrative burden and professional burnout. JAMA Netw Open. 2025;8(10):e2534976.

ONC (Office of the National Coordinator for Health Information Technology). 21st Century Cures Act: Interoperability, Information Blocking, and the ONC Health IT Certification Program. Washington, DC: ONC; 2020. Federal Register 85 FR 25642.

ONC (Office of the National Coordinator for Health Information Technology). 21st Century Cures Act of 2016. Pub. L. No. 114-255, 130 Stat. 1033. Title IV, § 4004, Information Blocking. Enacted December 13, 2016.

ONC (Office of the National Coordinator for Health Information Technology). 21st Century Cures Act Provider Disincentives Final Rule. Washington, DC: ONC; 2024.

ONC (Office of the National Coordinator for Health Information Technology). Audacious Inquiry. Patient Identification and Matching Final Report. Prepared for the Office of the National Coordinator for Health Information Technology (ONC). Washington, DC: ONC; 2014.

ONC (Office of the National Coordinator for Health Information Technology). Information Blocking Final Rule and Implementation. Washington, DC: ONC; 2023.

ONC (Office of the National Coordinator for Health Information Technology). Non-Federal Acute Care Hospital Electronic Health Record Adoption, 2008-2024. Washington, DC: ONC; 2026.

ONC (Office of the National Coordinator for Health Information Technology). Safety

Assurance Factors for EHR Resilience (SAFER) Guides. US Dept of Health and Human Services; 2025. Accessed August 31, 2026.

Osheroff JA, Teich JM, Levick D, et al. Improving Outcomes with CDS: An Implementer's Guide. 3rd ed. Chicago, IL: HIMSS; 2024.

Peterson-KFF Health System Tracker. Recent and Forecasted Trends in Prescription Drug Spending. San Francisco, CA: Peterson Center on Healthcare and KFF; 2025.

Pew Charitable Trusts. Enhanced Patient Matching Is Critical to Achieving Full Promise of Digital Health Records. Philadelphia, PA: Pew Charitable Trusts; 2018.

Phansalkar S, Van der Sijs H, Tucker AD, et al. Drug-drug interactions that should be non-interruptive to reduce alert fatigue in EHR. J Am Med Inform Assoc. 2013;20(3):489-493.

Pierson E, Cutler DM, Leskovec J, et al. An algorithmic approach to reducing unexplained pain disparities in underserved populations. Nat Med. 2021;27(1):136–140.

Pines JM, Hilton JA, Weber EJ, et al. International perspectives on ED crowding. Acad Emerg Med. 2011;18(12):1358-1370.

Premier, Inc. Claims Adjudication Costs Providers $25.7 Billion; $18 Billion Is Potentially Unnecessary Expense. Charlotte, NC: Premier, Inc.; February 24, 2025.

Procop GW, Yerian LM, Wyllie R, et al. Duplicate Laboratory Test Reduction Using a Clinical Decision Support Tool. American Journal of Clinical Pathology. 2014;141(5):718-723.

Rabin E, Kocher K, McClelland M, et al. Solutions to ED "boarding" and crowding are underused and may need to be legislated. Health Aff (Millwood). 2012;31(8):1757-1766.

Rabotin A, Aguilar E, Sandoval Gonzalez S, et al. Propagation of interpreter errors by ambient AI scribes: study using simulated clinical encounters. JMIR Medical Informatics. 2026;14:e88734.

Radley DC, Wasserman MR, Olsho LE, et al. Reduction in medication errors in hospitals due to adoption of CPOE systems. J Am Med Inform Assoc. 2013;20(3):470-476.

Ray KN, Shi Z, Gidengil CA, et al. Antibiotic prescribing during pediatric direct-to-consumer telemedicine visits. Pediatrics. 2019;143(5):e20182491.

Reason J. Human error: models and management. BMJ. 2000;320(7237):768-770.

Reinhardt, UE. "Divide et Impera: Protecting the Growth of Health Care Incomes (Costs)." 2012. Health Economics 21;1:41-54.

Reinhardt, UE. "What Doctors Make, and Why" (letter to the editor). New York Times Opinion section. 2007.

Rhee C, Chiotos K, Cosgrove SE, et al. Infectious Diseases Society of America position paper: recommended revisions to the national severe sepsis and septic shock early management bundle (SEP-1) sepsis quality measure. Clin Infect Dis. 2021;72(4):541-552.

Rhoades M, MacFeely S. Higher electronic health record functionality is associated with lower operating costs in urban but not rural hospitals. Applied Clinical Informatics. 2022;13(4):844-853.

Rosenbloom ST, Stead WW, Denny JC, et al. Generating clinical notes for EHR systems. Appl Clin Inform. 2010;1(3):232-243.

Rotenstein LS, Holmgren AJ, Downing NL, Bates DW. System-level factors and time spent on EHR by primary care physicians. JAMA Netw Open. 2023;6(11):e2344101.

Rotenstein LS, Holmgren AJ, Thombley R, et al. Changes in Clinician Time Expenditure and Visit Quantity With Adoption of Artificial Intelligence-Powered Scribes: A Multisite Study. JAMA. 2026.

Rule A, Bedrick S, Chiang MF, Hribar MR. Length and redundancy of outpatient progress notes across a decade at an academic medical center. JAMA Netw Open. 2021;4(7):e2115334.

Sackett DL, Rosenberg WM, Gray JA, et al. Evidence based medicine: what it is and what it isn't. BMJ. 1996;312(7023):71-72.

Sahni NR, Carrus B, Cutler DM. Administrative simplification and the potential for saving a quarter-trillion dollars in health care. JAMA. 2021;326(17):1677-1678.

Sahni N, Stein G, Zemmel R, Cutler DM. The Potential Impact of Artificial Intelligence on Healthcare Spending. NBER Working Paper No. 30857. National Bureau of Economic Research. 2023.

Santos CAQ, Conover C, Shehab N, et al. Electronic measurement of a clinical quality measure for inpatient hypoglycemic events: a multicenter validation study. Med Care. 2020;58(10):927-933.

Saposnik G, Redelmeier D, Ruff CC, Tobler PN. Cognitive biases associated with medical decisions: a systematic review. BMC Med Inform Decis Mak. 2016;16(1):138.

The Sequoia Project. A Framework for Cross-Organizational Patient Identity Management. Vienna, VA: The Sequoia Project; 2018.

Shanafelt TD, Dyrbye LN, Sinsky C, et al. Relationship between clerical burden and characteristics of the electronic environment with physician burnout and professional satisfaction. Mayo Clin Proc. 2016;91(7):836-848.

Shortliffe EH, Cimino JJ, eds. Biomedical Informatics: Computer Applications in Health Care and Biomedicine. 5th ed. Cham, Switzerland: Springer; 2021.

Shrank WH, Rogstad TL, Parekh N. Waste in the US health care system: estimated costs and potential for savings. JAMA. 2019;322(15):1501-1509.

Sieja A, Markley K, Pell J, Gonzalez C, et al. Optimization sprints: improving clinician satisfaction and teamwork by rapidly reducing electronic health record burden. Mayo Clin Proc. 2019;94(5):793-802.

Singer AJ, Thode HC Jr, Viccellio P, Pines JM. The association between length of ED boarding and mortality. Acad Emerg Med. 2011;18(12):1324-1329.

Singh H, Meyer AND, Thomas EJ. The frequency of diagnostic errors in outpatient care: estimations from three large observational studies involving US adult populations. BMJ Qual Saf. 2014;23(9):727-731.

Singh H, Sittig DF. Measuring and improving patient safety through health information technology: the Health IT Safety Framework. BMJ Qual Saf. 2016;25(4):226-232.

Singh H, Thomas EJ, Mani S, et al. Timely follow-up of abnormal diagnostic imaging test results in an outpatient setting: are electronic medical records achieving their potential? Arch Intern Med. 2009;169(17):1578-1586.

Singh R, Bapna M, Diab AR, et al. How AI is used in FDA-authorized medical devices: a taxonomy across 1,016 authorizations. npj Digit Med. 2025;8:388.

Sinsky C, Colligan L, Li L, et al. Allocation of physician time in ambulatory practice: a time and motion study in 4 specialties. Ann Intern Med. 2016;165(11):753-760.

Sittig DF, Singh H. A new sociotechnical model for studying health information technology in complex adaptive healthcare systems. Qual Saf Health Care. 2010;19(Suppl 3):i68-i74.

Sjoding MW, Dickson RP, Iwashyna TJ, Gay SE, Valley TS. Racial bias in pulse oximetry measurement. N Engl J Med. 2020;383(25):2477-2478.

Slack WV. The patient's right to decide. Lancet. 1977;2(8031):240.

Slight SP, Seger DL, Franz C, Wong A, Bates DW. The vulnerabilities of computerized physician order entry systems: a qualitative study. J Am Med Inform Assoc. 2016;23(2):311-316.

Starmer AJ, Spector ND, Srivastava R, et al. Changes in medical errors after implementation of a handoff program. N Engl J Med. 2014;371(19):1803-1812.

Steinkamp J, Kantrowitz JJ, Airan-Javia S. Prevalence and sources of duplicate information in the electronic medical record. JAMA Netw Open. 2022;5(9):e2233348.

Strathern M. 'Improving ratings': audit in the British University system. Eur Rev. 1997;5(3):305-321.

Tierney AA, Gayre G, Hoberman B, et al. A randomized clinical trial of two ambient AI scribes: measuring documentation efficiency and physician burnout. NEJM AI. 2025a.

Tierney AA, Gayre G, Hoberman B, et al. Ambient artificial intelligence scribes: learnings after 1 year and over 2.5 million uses. NEJM Catalyst. Published online March 31, 2025b.

TJC (The Joint Commission). Sentinel Event Alert 54: Safe Use of Health IT. Oakbrook Terrace, IL: The Joint Commission; 2015.

TJC (The Joint Commission). Sentinel Event Statistics: Root Causes by Event Type. Oakbrook Terrace, IL: The Joint Commission; 2017.

Tucker KL, Sheppard JP, Stevens R, et al. Self-monitoring of blood pressure in hypertension: a systematic review and individual patient data meta-analysis. PLoS Medicine. 2017;14(9):e1002389.

Turer RW, Levy BP, Hron JD, et al. An open letter arguing for closure of the practice pathway for clinical informatics medical subspecialty certification. Appl Clin Inform. 2022;13(1):301-303.

VA Office of Inspector General. Review of Alleged Patient Deaths, Patient Wait Times, and Scheduling Practices at the Phoenix VA Health Care System. Report No. 14-02603-267. Washington, DC: Department of Veterans Affairs; 2014.

Van der Sijs H, Aarts J, Vulto A, Berg M. Overriding of drug safety alerts in computerized physician order entry. J Am Med Inform Assoc. 2006;13(2):138-147.

Vest JR, Gamm LD. Health information exchange: persistent challenges and new strategies. J Am Med Inform Assoc. 2010;17(3):288-294.

Wadhera RK, Joynt Maddox KE, Wasfy JH, et al. Association of the Hospital Readmissions Reduction Program with mortality among Medicare beneficiaries hospitalized for heart failure, acute myocardial infarction, and pneumonia. JAMA. 2018;320(24):2542-2552.

Wanderer J, Mishra P, Ehrenfeld J. Innovation and market consolidation among electronic health record vendors: an acute need for regulation. J Med Syst. 2014;38:8.

Wang MD, Khanna R, Najafi N. Characterizing the source of text in electronic health record progress notes. JAMA Intern Med. 2017;177(8):1212-1213.

Washington V, DeSalvo K, Mostashari F, Blumenthal D. The HITECH era and the path forward. N Engl J Med. 2017;377(10):904-906.

Weed LL. Medical records that guide and teach. N Engl J Med. 1968;278(11):593-600, and 278(12):652-657.

Weis JM, Levy PC. Copy, paste, and cloned notes in EHR: prevalence, benefits, risks, and best practice recommendations. Chest. 2014;145(3):632-638.

Weiskopf NG, Weng C. Methods and dimensions of electronic health record data quality assessment: enabling reuse for clinical research. J Am Med Inform Assoc. 2013;20(1):144-151.

Wennberg JE. Tracking Medicine: A Researcher's Quest to Understand Health Care. Oxford, UK: Oxford University Press; 2010.

White AA, Shaver J, Melnick ER. Measuring implicit bias in ICU notes using word-embedding neural networks: Stability across time and geography. Chest. 2023;164(2):411-420.

WHO (World Health Organization). Global patient safety report 2024. World Health Organization; 2024.

WHO (World Health Organization). International Classification of Diseases, Eleventh Revision (ICD-11). Geneva: World Health Organization; 2022.

WittKieffer and Association of Medical Directors of Information Systems (AMDIS). The evolving role of the physician informaticist: Leadership, AI expansion, and structural governance. WittKieffer Healthcare Leadership Reports. 2025.

Wong A, Otles E, Donnelly JP, et al. External validation of a widely implemented proprietary sepsis prediction model in hospitalized patients. JAMA Intern Med. 2021;181(8):1065-1070.

Woolhandler S, Campbell T, Himmelstein DU. Costs of health care administration in the United States and Canada. N Engl J Med. 2003;349(8):768-775. doi:10.1056/NEJMsa022033.

Wright A, McEvoy DS, Aaron S, et al. Structured override reasons for drug-drug interaction alerts in electronic health records. J Am Med Inform Assoc. 2019;26(10):934-942.

ACKNOWLEDGMENTS

While this book's synthesis comes from my professional experience, the gravity is only granted by the excellent work of the many clinicians, authors, and scientists I referenced. One person saying something novel might be interesting or entertaining; decades of people saying the same things demand attention.

The largest debt is to people I cannot name easily. The clinical vignettes used are composites drawn from the patterns that emerged across many organizations over many years, relayed by many clinicians. Some were frustrated. Most of them were tired. All of them were generous with their time.

I want to specifically call out the scientific founders of informatics, data science and standards, and electronic health records. You will find many of these names in Chapter 17, and tied to key research in my bibliography, but the list is longer. They were all right early, they were specific, and they were published. This book is an attempt to review the notes they left.

One of them wrote a foreword to this book. Ed Hammond has spent five decades working on the problems this book refers upward in its final chapter, long before the money moved. That he read this manuscript and endorsed it means that I have done some justice to the scientific fundamentals, and this means a great deal to me. To David Lubarsky, who wrote the other foreword. He has run multiple high-functioning healthcare organizations of the kind this book describes as possible. He did it as a physician, and from the highest decision-making seat. His endorsement suggests that I have done some justice to the clinical and business case, which could not mean more to me. And, to Jacques Orces, who read it and endorsed it early. He is an executive-physician who experienced parallel perils, first as a CMIO, and now as a CMO. His endorsement means that I have done some justice to the breadth and depth of the career and informatics issues, which means my effort has been worthwhile.

To the analysts, builders, trainers, documentation specialists, and report writers I have worked beside for three decades. This workforce has stepped up many times to bridge the system's structural failures.

To the nurses and clinicians at bedside. Your efforts have not gone unnoticed. Many of the serious issues described in this book were heroically prevented from reaching a patient, by very hard work. I have not worked in an organization where that was not true.

To the physicians who still care a great deal about the care we deliver, and who let me follow them around, and take up valuable personal time.

To the many I have trained and mentored. I did it for a practical reason; I needed help. Many of them turned out to be excellent informaticists, which I am proudest of.

To my own discipline, which I have been hard on in these pages. My desire is that the pain causes action, which will in turn drive a new push for your skills.

To the executives who gave me the seat. And, to the ones who declined. This book exists because I saw what that choice costs.

ABOUT THE AUTHOR

Thomas Powell, MD MS, has served as chief medical information officer in academic, community, pediatric, and integrated delivery systems. In most of them, he was the first.

He went into medical informatics in the 1990s, before the work had a settled Name, Function, and recognized need. Three decades later he is still doing it, which puts him among a small group of people who have watched the full evolution.

He holds an MD from the University of Texas. He completed residency training at Duke University, where he also earned a master's degree in biomedical engineering with a concentration in clinical informatics.

Dr. Powell is the owner and medical director of CMIO, LLC, a national healthcare transformation practice. He offers interim chief medical information officer services to organizations that need the seat filled, and advises boards and executive teams on constituting the functions this book describes: the diagnostic assessment, the governance structure, the informatics workforce plan, and the disciplined measurement that follows. He also uses this Diagnostician's method to solve many of the targeted issues described in this book.

CMIO, LLC and the Diagnostician book website can be viewed at **thomaspowellmdcmio.com.** Numerous self-assessment tools, including the ones found in these Chapters and Appendices, are available for free.

He can be found at **linkedin.com/in/thomaspowell-md-cmio.**

www.ingramcontent.com/pod-product-compliance
Lightning Source LLC
LaVergne TN
LVHW051045210926
848486LV00001B/35

* 9 7 9 8 9 9 6 1 3 1 4 3 3 *